MW00910168

Cram101 Textbook Outlines to accompany:

Inquiry Into Physics

Vern J. Ostdiek, 6th Edition

A Cram101 Inc. publication (c) 2010.

PRACTICE EXAMS.

Get all of the self-teaching practice exams for each chapter of this textbook at **www.Cram101.com** and ace the tests. Here is an example:

Chapter 1

Inquiry Into Physics
Vern J. Ostdiek, 6th Edition,
All Material Written and Prepared by Cram101

I WANT A BETTER GRADE. Items 1 - 50 of 100.

1 In physics, and more specifically kinematics, _________ is the change in velocity over time. Because velocity is a vector, it can change in two ways: a change in magnitude and/or a change in direction. In one dimension, _________ is the rate at which something speeds up or slows down.

- Acceleration
- Ab initio multiple spawning
- Abbe number
- Abbe sine condition

2 _________ is the branch of physics concerned with the behaviour of physical bodies when subjected to forces or displacements, and the subsequent effect of the bodies on their environment. The discipline has its roots in several ancient civilizations During the early modern period, scientists such as Galileo, Kepler, and especially Newton, laid the foundation for what is now known as classical _________

- Mechanics
- M. Riesz extension theorem
- Ma Yize
- Macdonald polynomials

3 _________ is a concept used in the physical sciences to explain a number of observable behaviours, and in everyday usage, it is common to identify _________ with those resulting behaviors. In particular, _________ is commonly identified with weight. But according to our modern scientific understanding, the weight of an object results from the interaction of its _________ with a gravitational field, so while _________ is part of the explanation of weight, it is not the complete explanation.

- Mass
- Magi
- Mangal Dosha
- Mars effect

With Cram101.com online, you also have access to extensive reference material.

You will nail those essays and papers. Here is an example from a Cram101 Biology text:

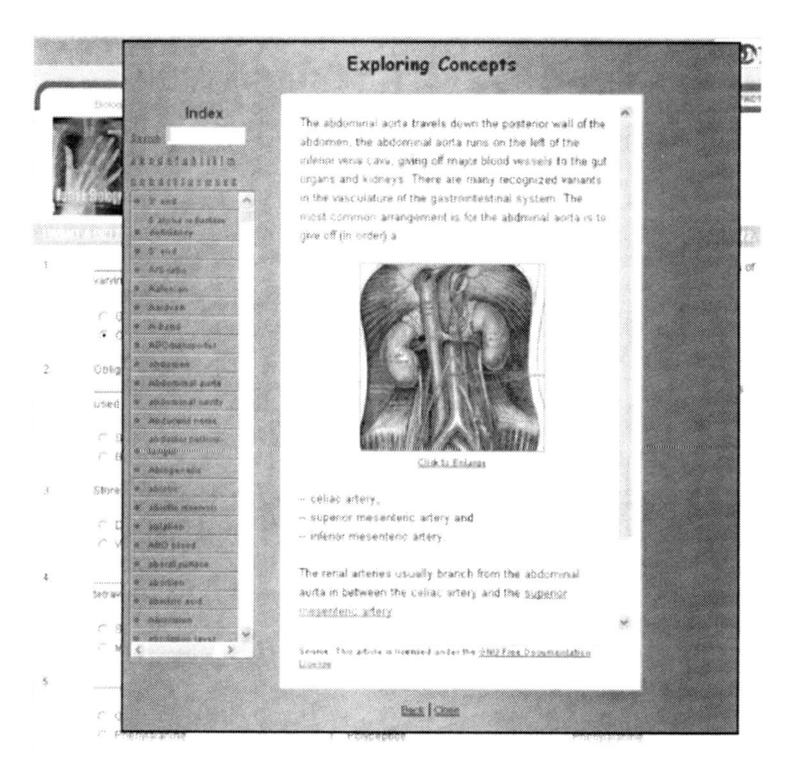

Visit **www.Cram101.com**, click Sign Up at the top of the screen, and enter DK73DW8448 in the promo code box on the registration screen. Access to www.Cram101.com is normally $9.95 per month, but because you have purchased this book, your access fee is only $4.95 per month, cancel at any time. Sign up and stop highlighting textbooks forever.

Learning System

Cram101 Textbook Outlines is a learning system. The notes in this book are the highlights of your textbook, you will never have to highlight a book again.

How to use this book. Take this book to class, it is your notebook for the lecture. The notes and highlights on the left hand side of the pages follow the outline and order of the textbook. All you have to do is follow along while your instructor presents the lecture. Circle the items emphasized in class and add other important information on the right side. With Cram101 Textbook Outlines you'll spend less time writing and more time listening. Learning becomes more efficient.

Cram101.com Online

Increase your studying efficiency by using Cram101.com's practice tests and online reference material. It is the perfect complement to Cram101 Textbook Outlines. Use self-teaching matching tests or simulate in-class testing with comprehensive multiple choice tests, or simply use Cram's true and false tests for quick review. Cram101.com even allows you to enter your in-class notes for an integrated studying format combining the textbook notes with your class notes.

Visit **www.Cram101.com**, click Sign Up at the top of the screen, and enter **DK73DW8448** in the promo code box on the registration screen. Access to www.Cram101.com is normally $9.95 per month, but because you have purchased this book, your access fee is only $4.95 per month. Sign up and stop highlighting textbooks forever.

 ISBN(s): 9781428897823.

Inquiry Into Physics
Vern J. Ostdiek, 6th

CONTENTS

Acceleration	In physics, and more specifically kinematics, Acceleration is the change in velocity over time. Because velocity is a vector, it can change in two ways: a change in magnitude and/or a change in direction. In one dimension, Acceleration is the rate at which something speeds up or slows down.
Mechanics	Mechanics is the branch of physics concerned with the behaviour of physical bodies when subjected to forces or displacements, and the subsequent effect of the bodies on their environment. The discipline has its roots in several ancient civilizations During the early modern period, scientists such as Galileo, Kepler, and especially Newton, laid the foundation for what is now known as classical Mechanics
Mass	Mass is a concept used in the physical sciences to explain a number of observable behaviours, and in everyday usage, it is common to identify Mass with those resulting behaviors. In particular, Mass is commonly identified with weight. But according to our modern scientific understanding, the weight of an object results from the interaction of its Mass with a gravitational field, so while Mass is part of the explanation of weight, it is not the complete explanation.
Matter	The term Matter traditionally refers to the substance that objects are made of. One common way to identify this 'substance' is through its physical properties; a common definition of Matter is anything that has mass and occupies a volume. However, this definition has to be revised in light of quantum mechanics, where the concept of 'having mass', and 'occupying space' are not as well-defined as in everyday life.
Physical quantity	A physical quantity is a physical property that can be quantified. This means it can be measured and/or calculated and expressed in numbers. The value of a physical quantity Q is expressed as the product of a numerical value {Q} and a physical unit .
Space	Space is the boundless, three-dimensional extent in which objects and events occur and have relative position and direction. Physical Space is often conceived in three linear dimensions, although modern physicists usually consider it, with time, to be part of the boundless four-dimensional continuum known as Space time. In mathematics Space s with different numbers of dimensions and with different underlying structures can be examined.
Universe	The Universe is defined as everything that physically exists: the entirety of space and time, all forms of matter, energy and momentum, and the physical laws and constants that govern them. However, the term 'universe' may be used in slightly different contextual senses, denoting such concepts as the cosmos, the world or Nature. Astronomical observations indicate that the universe is 13.73 ± 0.12 billion years old and at least 93 billion light years across.
Area	Area is a quantity expressing the two-dimensional size of a defined part of a surface, typically a region bounded by a closed curve. The term surface Area refers to the total Area of the exposed surface of a 3-dimensional solid, such as the sum of the Area s of the exposed sides of a polyhedron. Area is an important invariant in the differential geometry of surfaces.
Volume	The Volume of any solid, liquid, plasma, vacuum or theoretical object is how much three-dimensional space it occupies, often quantified numerically. One-dimensional figures (such as lines) and two-dimensional shapes such as square geometry squares are assigned zero Volume in the three-dimensional space. Volume is commonly presented in units such as mililitres or cm^3 (milliliters or cubic centimeters.)

Cycle

In combinatorial mathematics, a cycle of length n of a permutation P over a set S is a subset { c_1, ..., c_n } of S on which the permutation P acts in the following way:

$P(c_i) = c_{i+1}$ for i = 1, ..., n − 1, and $P(c_n) = c_1$.

It is usual to write a cycle by its mapping:

$(c_1 \, c_2 \ldots c_n \, .)$

Earth

Earth is the third planet from the Sun. It is the fifth largest of the eight planets in the solar system, and the largest of the terrestrial planets (non-gas planets) in the Solar System in terms of diameter, mass and density. It is also referred to as the World, the Blue Planet, and Terra.

Pendulum

A pendulum is a mass that is attached to a pivot, from which it can swing freely. This object is subject to a restoring force due to gravity that will accelerate it toward an equilibrium position. When the pendulum is displaced from its place of rest, the restoring force will cause the pendulum to oscillate about the equilibrium position.

Period

In the periodic table of the elements, a Period is a horizontal row of the table. A group, or family, is a vertical column of the table.

The elements are arranged in a series of rows so that those with similar properties appear in vertical columns.

Second

The second (SI symbol: s), sometimes abbreviated sec., is the name of a unit of time, and is the International System of Units (SI) base unit of time. It may be measured using a clock.

SI prefixes are frequently combined with the word second to denote subdivisions of the second e.g., the milli second (one thousandth of a second , the micro second (one millionth of a second , and the nano second (one billionth of a second)

Ephemeris

An Ephemeris is a table of values that gives the positions of astronomical objects in the sky at a given time or times. Different kinds are used for astronomy and astrology. Even though this was also one of the first applications of mechanical computers, an Ephemeris will still often be a simple printed table.

Expected value

In probability theory and statistics, the expected value (or expectation value or mean, or first moment) of a random variable is the integral of the random variable with respect to its probability measure.

For discrete random variables this is equivalent to the probability-weighted sum of the possible values.

For continuous random variables with a density function it is the probability density-weighted integral of the possible values.

Satellite

In the context of spaceflight, a satellite is an object which has been placed into orbit by human endeavor. Such objects are sometimes called artificial satellite s to distinguish them from natural satellite s such as the Moon.

The first artificial satellite Sputnik 1, was launched by the Soviet Union in 1957.

Universal gravitation

Newton's law of universal gravitation is a physical law describing the gravitational attraction between bodies with mass. It is a part of classical mechanics and was first formulated in Newton's work Philosophiae Naturalis Principia Mathematica, first published on July 5, 1687. In modern language it states the following:

Every point mass attracts every other point mass by a force pointing along the line intersecting both points.

Atom

The Atom is a basic unit of matter consisting of a dense, central nucleus surrounded by a cloud of negatively charged electrons. The Atom ic nucleus contains a mix of positively charged protons and electrically neutral neutrons (except in the case of hydrogen-1, which is the only stable nuclide with no neutron.) The electrons of an Atom are bound to the nucleus by the electromagnetic force.

Conservation law

In physics, a Conservation law states that a particular measurable property of an isolated physical system does not change as the system evolves.

One particularly important physical result concerning Conservation law s is Noether's Theorem, which states that there is a one-to-one correspondence between Conservation law s and differentiable symmetries of physical systems. For example, the conservation of energy follows from the time-invariance of physical systems, and the fact that physical systems behave the same regardless of how they are oriented in space gives rise to the conservation of angular momentum.

Measurement

The framework of quantum mechanics requires a careful definition of measurement The issue of measurement lies at the heart of the problem of the interpretation of quantum mechanics, for which there is currently no consensus.

Solar

Solar power uses Solar Radiation emitted from our sun. Solar power, a renewable energy source, has been used in many traditional technologies for centuries, and is in widespread use where other power supplies are absent, such as in remote locations and in space.

Solar day

Apparent solar time or true solar time is the hour angle of the Sun. It is based on the apparent solar day, which is the interval between two successive returns of the Sun to the local meridian. Note that the solar day starts at noon, so apparent solar time 00:00 means noon and 12:00 means midnight.

Frequency

Frequency is the number of occurrences of a repeating event per unit time. It is also referred to as temporal Frequency The period is the duration of one cycle in a repeating event, so the period is the reciprocal of the Frequency

Radio

Radio is the transmission of signals, by modulation of electromagnetic waves with frequencies below those of visible light. Electromagnetic radiation travels by means of oscillating electromagnetic fields that pass through the air and the vacuum of space. Information is carried by systematically changing some property of the radiated waves, such as amplitude, frequency, or phase.

Radio frequency

Radio frequency is a frequency or rate of oscillation within the range of about 3 Hz to 300 GHz. This range corresponds to frequency of alternating current electrical signals used to produce and detect radio waves. Since most of this range is beyond the vibration rate that most mechanical systems can respond to, RF usually refers to oscillations in electrical circuits or electromagnetic radiation.

Inertia

Inertia is the resistance of mass, i.e. any physical object, to a change in its state of motion. The principle of Inertia is one of the fundamental principles of classical physics which are used to describe the motion of matter and how it is affected by applied forces. Inertia comes from the Latin word, 'iners', meaning idle, or lazy.

Specular reflection

Specular reflection is the perfect, mirror-like reflection of light from a surface, in which light from a single incoming direction is reflected into a single outgoing direction. Such behavior is described by the law of reflection, which states that the direction of incoming light, and the direction of outgoing light reflected make the same angle with respect to the surface normal, thus the angle of incidence equals the angle of reflection; this is commonly stated as $>\theta_i = >\theta_r$. This behavior was first discovered through careful observation and measurement by Hero of Alexandria.>

This is in contrast to diffuse reflection, where incoming light is reflected in a broad range of directions.

Lightning

Lightning is an atmospheric discharge of electricity accompanied by thunder, which typically occurs during thunderstorms, and sometimes during volcanic eruptions or dust storms. In the atmospheric electrical discharge, a leader of a bolt of Lightning can travel at speeds of 60,000 m/s (130,000 mph), and can reach temperatures approaching 30,000 °C (54,000 °F), hot enough to fuse silica sand into glass channels known as fulgurites which are normally hollow and can extend some distance into the ground. There are some 16 million Lightning storms in the world every year.

Vector

In elementary mathematics, physics, and engineering, a vector is a geometric object that has both a magnitude and a direction. A vector is frequently represented by a line segment with a definite direction, or graphically as an arrow, connecting an initial point A with a terminal point B, and denoted by

$$\overrightarrow{AB}.$$

The magnitude of the vector is the length of the segment and the direction characterizes the displacement of B relative to A: how much one should move the point A to 'carry' it to the point B.

Many algebraic operations on real numbers have close analogues for vectors.

Velocity

In physics, velocity is defined as the rate of change of position. It is a vector physical quantity; both speed and direction are required to define it. In the SI system, it is measured in meters per second: or ms^{-1}.

Slope

Slope is used to describe the steepness, incline, gradient, or grade of a straight line. A higher slope value indicates a steeper incline. The slope is defined as the ratio of the 'rise' divided by the 'run' between two points on a line, or in other words, the ratio of the altitude change to the horizontal distance between any two points on the line.

Air

The Earth's atmosphere is a layer of gases surrounding the planet Earth that is retained by Earth's gravity. The atmosphere protects life on Earth by absorbing ultraviolet solar radiation, warming the surface through heat retention (greenhouse effect), and reducing temperature extremes between day and night. Dry Air contains roughly (by volume) 78.08% nitrogen, 20.95% oxygen, 0.93% argon, 0.038% carbon dioxide, and trace amounts of other gases.

Quantum

In physics, a quantum is an indivisible entity of a quantity that has the same units as the Planck constant and is related to both energy and momentum of elementary particles of matter and of photons and other bosons. The word comes from the Latin 'quantus,' for 'how much.' Behind this, one finds the fundamental notion that a physical property may be 'quantized', referred to as 'quantization'. This means that the magnitude can take on only certain discrete numerical values, rather than any value, at least within a range.

Quantum mechanics

Quantum mechanics is the study of mechanical systems whose dimensions are close to the atomic scale, such as molecules, atoms, electrons, protons and other subatomic particles. Quantum mechanics is a fundamental branch of physics with wide applications. Quantum theory generalizes classical mechanics to provide accurate descriptions for many previously unexplained phenomena such as black body radiation and stable electron orbits.

Pythagoras

Pythagoras of Samos was an Ionian Greek mathematician and founder of the religious movement called Pythagoreanism. He is often revered as a great mathematician, mystic and scientist; however some have questioned the scope of his contributions to mathematics and natural philosophy. Herodotus referred to him as 'the most able philosopher among the Greeks'.

Terminal

A Terminal is a conductive device for joining electrical circuits together. The connection may be temporary, as for portable equipment, or may require a tool for assembly and removal, or may be a permanent electrical joint between two wires or devices.

Terminals form a single circuit, unlike electrical connectors, which form two or more circuits.

Terminal velocity

A free falling object achieves its terminal velocity when the downward force of gravityequals the upward force of drag. This causes the net force on the object to be zero, resulting in an acceleration of zero. Mathematically an object asymptotically approaches and can never reach its terminal velocity.

Observation

Observation is either an activity of a living being (such as a human), consisting of receiving knowledge of the outside world through the senses, or the recording of data using scientific instruments. The term may also refer to any datum collected during this activity.

The scientific method requires Observation s of nature to formulate and test hypotheses.

Physical science

Physical science is an encompassing term for the branches of natural science and science that study non-living systems, in contrast to the biological sciences. However, the term 'physical' creates an unintended, somewhat arbitrary distinction, since many branches of Physical science also study biological phenomena.

The following outline is presented as an overview of and topical guide to Physical science

The foundations of the Physical science s rests upon key concepts and theories, each of which explains and/or models a particular aspect of the behavior of nature.

Angular momentum

Angular momentum is a quantity that is useful in describing the rotational state of a physical system. For a rigid body rotating around an axis of symmetry (e.g. the fins of a ceiling fan), the Angular momentum can be expressed as the product of the body's moment of inertia and its angular velocity ($\mathbf{L} = I\boldsymbol{\omega}$.) In this way, Angular momentum is sometimes described as the rotational analog of linear momentum.

Linear

The word Linear comes from the Latin word Linear is, which means created by lines. In mathematics, a Linear map or function f(x) is a function which satisfies the following two properties...

· Additivity (also called the superposition property): $f(x + y) = f(x) + f(y.)$

Angular momentum | Angular momentum is a quantity that is useful in describing the rotational state of a physical system. For a rigid body rotating around an axis of symmetry (e.g. the fins of a ceiling fan), the Angular momentum can be expressed as the product of the body's moment of inertia and its angular velocity ($\mathbf{L} = I\boldsymbol{\omega}$.) In this way, Angular momentum is sometimes described as the rotational analog of linear momentum.

Force | In physics, a Force is any external agent that causes a change in the motion of a free body, or that causes stress in a fixed body. It can also be described by intuitive concepts such as a push or pull that can cause an object with mass to change its velocity , i.e., to accelerate, or which can cause a flexible object to deform. Force has both magnitude and direction, making it a vector quantity.

Horizon | The Horizon is the apparent line that separates earth from sky.

It is the line that divides all visible directions into two categories: those that intersect the Earth's surface, and those that do not. At many locations, the true Horizon is obscured by trees, buildings, mountains, etc., and the resulting intersection of earth and sky is called the visible Horizon

Linear | The word Linear comes from the Latin word Linear is, which means created by lines. In mathematics, a Linear map or function f(x) is a function which satisfies the following two properties...

· Additivity (also called the superposition property): f(x + y) = f(x) + f(y.)

Mechanics | Mechanics is the branch of physics concerned with the behaviour of physical bodies when subjected to forces or displacements, and the subsequent effect of the bodies on their environment. The discipline has its roots in several ancient civilizations During the early modern period, scientists such as Galileo, Kepler, and especially Newton, laid the foundation for what is now known as classical Mechanics

Spacecraft | A spacecraft is a vehicle or machine designed for spaceflight. On a sub-orbital spaceflight, a spacecraft enters outer space but then returns to the planetary surface without making a complete orbit. For an orbital spaceflight, a spacecraft enters a closed orbit around the planetary body.

Universal gravitation | Newton's law of universal gravitation is a physical law describing the gravitational attraction between bodies with mass. It is a part of classical mechanics and was first formulated in Newton's work Philosophiae Naturalis Principia Mathematica, first published on July 5, 1687. In modern language it states the following:

Every point mass attracts every other point mass by a force pointing along the line intersecting both points.

Conservation law | In physics, a Conservation law states that a particular measurable property of an isolated physical system does not change as the system evolves.

One particularly important physical result concerning Conservation law s is Noether's Theorem, which states that there is a one-to-one correspondence between Conservation law s and differentiable symmetries of physical systems. For example, the conservation of energy follows from the time-invariance of physical systems, and the fact that physical systems behave the same regardless of how they are oriented in space gives rise to the conservation of angular momentum.

Energy

In physics, Energy is a scalar physical quantity that describes the amount of work that can be performed by a force, an attribute of objects and systems that is subject to a conservation law. Different forms of Energy include kinetic, potential, thermal, gravitational, sound, light, elastic, and electromagnetic Energy The forms of Energy are often named after a related force.

Physical science

Physical science is an encompassing term for the branches of natural science and science that study non-living systems, in contrast to the biological sciences. However, the term 'physical' creates an unintended, somewhat arbitrary distinction, since many branches of Physical science also study biological phenomena.

The following outline is presented as an overview of and topical guide to Physical science

The foundations of the Physical science s rests upon key concepts and theories, each of which explains and/or models a particular aspect of the behavior of nature.

Vector

In elementary mathematics, physics, and engineering, a vector is a geometric object that has both a magnitude and a direction. A vector is frequently represented by a line segment with a definite direction, or graphically as an arrow, connecting an initial point A with a terminal point B, and denoted by

The magnitude of the vector is the length of the segment and the direction characterizes the displacement of B relative to A: how much one should move the point A to 'carry' it to the point B.

Many algebraic operations on real numbers have close analogues for vectors.

Velocity

In physics, velocity is defined as the rate of change of position. It is a vector physical quantity; both speed and direction are required to define it. In the SI system, it is measured in meters per second: or ms^{-1}.

Static

Statics is the branch of mechanics concerned with the analysis of loads on physical systems in static equilibrium, that is, in a state where the relative positions of subsystems do not vary over time, or where components and structures are at rest. When in static equilibrium, the system is either at rest, or its center of mass moves at constant velocity.

By Newton's second law, this situation implies that the net force and net torque on every body in the system is zero, meaning that for every force bearing upon a member, there must be an equal and opposite force.

Static friction

Friction is the force resisting the relative lateral (tangential) motion of solid surfaces, fluid layers, or material elements in contact. It is usually subdivided into several varieties:

· Dry friction resists relative lateral motion of two solid surfaces in contact. Dry friction is also subdivided into Static friction between non-moving surfaces, and kinetic friction (sometimes called sliding friction or dynamic friction) between moving surfaces.

· Lubricated friction or fluid friction resists relative lateral motion of two solid surfaces separated by a layer of gas or liquid.

· Fluid friction is also used to describe the friction between layers within a fluid that are moving relative to each other.

· Skin friction is a component of drag, the force resisting the motion of a solid body through a fluid.

· Internal friction is the force resisting motion between the elements making up a solid material while it undergoes deformation.

Friction is not a fundamental force, as it is derived from electromagnetic force between charged particles, including electrons, protons, atoms, and molecules, and so cannot be calculated from first principles, but instead must be found empirically. When contacting surfaces move relative to each other, the friction between the two surfaces converts kinetic energy into thermal energy, or heat.

Air

The Earth's atmosphere is a layer of gases surrounding the planet Earth that is retained by Earth's gravity. The atmosphere protects life on Earth by absorbing ultraviolet solar radiation, warming the surface through heat retention (greenhouse effect), and reducing temperature extremes between day and night. Dry Air contains roughly (by volume) 78.08% nitrogen, 20.95% oxygen, 0.93% argon, 0.038% carbon dioxide, and trace amounts of other gases.

Net force

A Net force $F_{net} = F_1 + F_2 +$ â€¦ (also known as a resultant force) is a vector produced when two or more forces { F_1, F_2, â€¦ } act upon a single object. It is calculated by vector addition of the force vectors acting upon the object. A Net force can also be defined as the overall force acting on an object, when all the individual forces acting on the object are added together.

Tribology

Tribology is the science and technology of interacting surfaces in relative motion. It includes the study and application of the principles of friction, lubrication and wear. The word 'tribology' derives from the Greek τρÎ¯βω meaning rub', and λÏŒγος meaning 'principle or logic'.

Wave

A Wave is a disturbance that propagates through space and time, usually with transference of energy. A mechanical Wave is a Wave that propagates or travels through a medium due to the restoring forces it produces upon deformation. There also exist Wave s capable of traveling through a vacuum, including electromagnetic radiation and probably gravitational radiation.

Waves

Waves is a three-part novel by Ogan Gurel published in 2009. A 21st century version of Faust, the novel explores good and evil in both individual and global settings focusing around a hypothetical technology that has both medical and military applications. The protagonist, Tomas Twarok, is a contemplative and idealistic doctor-turned-entrepreneur who strikes a deal with his college friend, Maximilian Iblis, a ruthless hedge fund manager.

Biomechanics

Biomechanics is the science concerned with the action of forces, internal or external, on the living body . Biological masses are under the influence of both intrinsic and extrinsic forces of gravity. Comaparative impact of extrinsic forces of Earth's gravity as well as of micro-gravity, as experienced in space, are being studied on living organisms.

Nanotribology

Nanotribology is a branch of tribology which studies friction phenomenon at the nanometer scale The distinction between Nanotribology and tribology is primarily due to the involvement of atomic forces in the determination of the final behavior of the system.

Gears, bearings, and liquid lubricants can reduce friction in the macroscopic world, but the origins of friction for small devices such as micro- or nanoelectromechanical systems (NEMS) require other solutions.

Irreversibility	In science, a process that is not reversible is called irreversible. This concept arises most frequently in thermodynamics, as applied to processes. Irreversibility is also used in economics to refer to investment or expenditures that involve large sunk costs.
Acceleration	In physics, and more specifically kinematics, Acceleration is the change in velocity over time. Because velocity is a vector, it can change in two ways: a change in magnitude and/or a change in direction. In one dimension, Acceleration is the rate at which something speeds up or slows down.
Centripetal force	Centripetal force is a force that makes a body follow a curved, as opposed to straight, path; it is always directed orthogonal to the velocity of the body, toward the instantaneous center of curvature of the path. The term Centripetal force comes from the Latin words centrum and petere ('tend towards', 'aim at'), signifying that the force is directed inward toward the center of curvature of the path. Isaac Newton's description was: 'A Centripetal force is that by which bodies are drawn or impelled, or in any way tend, towards a point as to a center.' and is found in the Principia.
Space	Space is the boundless, three-dimensional extent in which objects and events occur and have relative position and direction. Physical Space is often conceived in three linear dimensions, although modern physicists usually consider it, with time, to be part of the boundless four-dimensional continuum known as Space time. In mathematics Space s with different numbers of dimensions and with different underlying structures can be examined.
Space station	A space station is an artificial structure designed for humans to live in outer space. So far only low earth orbit stations are implemented, also known as orbital stations. A space station is distinguished from other manned spacecraft by its lack of major propulsion or landing facilities -- instead, other vehicles are used as transport to and from the station.
Inertia	Inertia is the resistance of mass, i.e. any physical object, to a change in its state of motion. The principle of Inertia is one of the fundamental principles of classical physics which are used to describe the motion of matter and how it is affected by applied forces. Inertia comes from the Latin word, 'iners', meaning idle, or lazy.
Mass	Mass is a concept used in the physical sciences to explain a number of observable behaviours, and in everyday usage, it is common to identify Mass with those resulting behaviors. In particular, Mass is commonly identified with weight. But according to our modern scientific understanding, the weight of an object results from the interaction of its Mass with a gravitational field, so while Mass is part of the explanation of weight, it is not the complete explanation.
Second	The second (SI symbol: s), sometimes abbreviated sec., is the name of a unit of time, and is the International System of Units (SI) base unit of time. It may be measured using a clock. SI prefixes are frequently combined with the word second to denote subdivisions of the second e.g., the milli second (one thousandth of a second , the micro second (one millionth of a second , and the nano second (one billionth of a second)
Matter	The term Matter traditionally refers to the substance that objects are made of. One common way to identify this 'substance' is through its physical properties; a common definition of Matter is anything that has mass and occupies a volume. However, this definition has to be revised in light of quantum mechanics, where the concept of 'having mass', and 'occupying space' are not as well-defined as in everyday life.

Subatomic	A subatomic particle is an elementary or composite particle smaller than an atom. Particle physics and nuclear physics are concerned with the study of these particles, their interactions, and non-atomic matter. subatomic particles include the atomic constituents electrons, protons, and neutrons.
Subatomic particle	A subatomic particle is an elementary or composite particle smaller than an atom. Particle physics and nuclear physics are concerned with the study of these particles, their interactions, and non-atomic matter. Subatomic particles include the atomic constituents electrons, protons, and neutrons.
Subatomic particles	In physics, subatomic particles are the particles composing nucleons and atoms. There are two types of subatomic particles elementary particles, which are not made of other particles, and composite particles. Particle physics and nuclear physics study these particles and how they interact.
Volume	The Volume of any solid, liquid, plasma, vacuum or theoretical object is how much three-dimensional space it occupies, often quantified numerically. One-dimensional figures (such as lines) and two-dimensional shapes such as square geometry squares are assigned zero Volume in the three-dimensional space. Volume is commonly presented in units such as mililitres or cm^3 (milliliters or cubic centimeters.)
Measurement	The framework of quantum mechanics requires a careful definition of measurement The issue of measurement lies at the heart of the problem of the interpretation of quantum mechanics, for which there is currently no consensus.
Parabola	In mathematics, the parabola is a conic section, the intersection of a right circular conical surface and a plane parallel to a generating straight line of that surface. Given a point and a line that lie in a plane, the locus of points in that plane that are equidistant to them is a parabola. A particular case arises when the plane is tangent to the conical surface of a circle.
Trajectory	Trajectory is the path a moving object follows through space. The object might be a projectile or a satellite, for example. It thus includes the meaning of orbit - the path of a planet, an asteroid or a comet as it travels around a central mass.
Harmonic	In acoustics and telecommunication, a Harmonic of a wave is a component frequency of the signal that is an integer multiple of the fundamental frequency. For example, if the fundamental frequency is f, the Harmonic s have frequencies f, 2f, 3f, 4f, etc. The Harmonic s have the property that they are all periodic at the fundamental frequency, therefore the sum of Harmonic s is also periodic at that frequency.
Simple harmonic motion	Simple harmonic motion is the motion of a simple harmonic oscillator, a motion that is neither driven nor damped. The motion is periodic, as it repeats itself at standard intervals in a specific manner - described as being sinusoidal, with constant amplitude. It is characterized by its amplitude, its period which is the time for a single oscillation, its frequency which is the number of cycles per unit time, and its phase, which determines the starting point on the sine wave.
Cycle	In combinatorial mathematics, a cycle of length n of a permutation P over a set S is a subset $\{ c_1, ..., c_n \}$ of S on which the permutation P acts in the following way: $P(c_i) = c_{i+1}$ for $i = 1, ..., n - 1$, and $P(c_n) = c_1$.

It is usual to write a cycle by its mapping:

$(c_1 \, c_2 \ldots c_n)$

Frequency

Frequency is the number of occurrences of a repeating event per unit time. It is also referred to as temporal Frequency
The period is the duration of one cycle in a repeating event, so the period is the reciprocal of the Frequency

Sine wave

The sine wave or sinusoid is a function that occurs often in mathematics, physics, signal processing, audition, electrical engineering, and many other fields. Its most basic form is:

$$y(t) = A \cdot \sin(\omega t + \theta)$$

which describes a wavelike function of time with:

· peak deviation from center = A
· angular frequency ω
· phase = θ

· When the phase is non-zero, the entire waveform appears to be shifted in time by the amount θ/ω seconds. A negative value represents a delay, and a positive value represents a 'head-start'.
The sine wave is important in physics because it retains its waveshape when added to another sine wave of the same frequency and arbitrary phase. It is the only periodic waveform that has this property. This property leads to its importance in Fourier analysis and makes it acoustically unique.

Shape

The shape of an object located in some space is the part of that space occupied by the object, as determined by its external boundary - abstracting from other properties such as colour, content, and material composition, as well as from the object's other spatial properties .
Mathematician and statistician David George Kendall defined shape this way:

Simple two-dimensional shape s can be described by basic geometry such as points, line, curves, plane, and so on.

Orbit

In physics, an orbit is the gravitationally curved path of one object around a point or another body, for example the gravitational orbit of a planet around a star.
Historically, the apparent motion of the planets were first understood in terms of epicycles, which are the sums of numerous circular motions. This predicted the path of the planets quite well, until Johannes Kepler was able to show that the motion of the planets were in fact elliptical motions.

Orbits

The velocity relationship of two objects with mass can thus be considered in four practical classes, with subtypes:

· No orbit
· Interrupted orbits

· Range of interrupted elliptical paths
· Circumnavigating orbits

· Range of elliptical paths with closest point opposite firing point
· Circular path
· Range of elliptical paths with closest point at firing point
· Infinite orbits

· Parabolic paths
· Hyperbolic paths

In many situations relativistic effects can be neglected, and Newton's laws give a highly accurate description of the motion. Then the acceleration of each body is equal to the sum of the gravitational forces on it, divided by its mass, and the gravitational force between each pair of bodies is proportional to the product of their masses and decreases inversely with the square of the distance between them. To this Newtonian approximation, for a system of two point masses or spherical bodies, only influenced by their mutual gravitation, the orbits can be exactly calculated. If the heavier body is much more massive than the smaller, as for a satellite or small moon orbiting a planet or for the Earth orbiting the Sun, it is accurate and convenient to describe the motion in a coordinate system that is centered on the heavier body, and we can say that the lighter body is in orbit around the heavier.

Space shuttle

NASA's Space Shuttle, officially called the Space Transportation System, is the spacecraft currently used by the United States government for its human spaceflight missions. At launch, it consists of a rust-colored external tank, two white, slender Solid Rocket Boosters, and the orbiter, a winged spaceplane which is the space shuttle in the narrow sense.

The orbiter carries astronauts and payload such as satellites or space station parts into low earth orbit, into the Earth's upper atmosphere or thermosphere.

Terminal

A Terminal is a conductive device for joining electrical circuits together. The connection may be temporary, as for portable equipment, or may require a tool for assembly and removal, or may be a permanent electrical joint between two wires or devices.

Terminals form a single circuit, unlike electrical connectors, which form two or more circuits.

Terminal velocity

A free falling object achieves its terminal velocity when the downward force of gravityequals the upward force of drag. This causes the net force on the object to be zero, resulting in an acceleration of zero. Mathematically an object asymptotically approaches and can never reach its terminal velocity.

Chaos

Chaos typically refers to a state lacking order or predictability. In ancient Greece, it referred to the initial state of the universe, and, by extension, space, darkness, or an abyss (the antithetical concept was cosmos.) In modern English, it is used in classical studies with this original meaning; in mathematics and science to refer to a very specific kind of unpredictability; and informally to mean a state of confusion.

Quantum	In physics, a quantum is an indivisible entity of a quantity that has the same units as the Planck constant and is related to both energy and momentum of elementary particles of matter and of photons and other bosons. The word comes from the Latin 'quantus,' for 'how much.' Behind this, one finds the fundamental notion that a physical property may be 'quantized', referred to as 'quantization'. This means that the magnitude can take on only certain discrete numerical values, rather than any value, at least within a range.
Quantum mechanics	Quantum mechanics is the study of mechanical systems whose dimensions are close to the atomic scale, such as molecules, atoms, electrons, protons and other subatomic particles. Quantum mechanics is a fundamental branch of physics with wide applications. Quantum theory generalizes classical mechanics to provide accurate descriptions for many previously unexplained phenomena such as black body radiation and stable electron orbits.
Randomness	Randomness is a lack of order, purpose, cause, or predictability. A random process is a repeating process whose outcomes follow no describable deterministic pattern, but follow a probability distribution. The term is often used in statistics to signify well-defined statistical properties, such as a lack of bias or correlation.
Earth	Earth is the third planet from the Sun. It is the fifth largest of the eight planets in the solar system, and the largest of the terrestrial planets (non-gas planets) in the Solar System in terms of diameter, mass and density. It is also referred to as the World, the Blue Planet, and Terra.
Force field	A Force field sometimes known as an energy shield, force shield typically made of energy or charged particles, that protects a person, area or object from attacks or intrusions. Force field s tend to appear often in works of speculative fiction. A University of Washington in Seattle group has been experimenting with using a bubble of charged plasma to surround a spacecraft, contained by a fine mesh of superconducting wire.
Atom	The Atom is a basic unit of matter consisting of a dense, central nucleus surrounded by a cloud of negatively charged electrons. The Atom ic nucleus contains a mix of positively charged protons and electrically neutral neutrons (except in the case of hydrogen-1, which is the only stable nuclide with no neutron.) The electrons of an Atom are bound to the nucleus by the electromagnetic force.
Molecule	A Molecule is defined as a sufficiently stable, electrically neutral group of at least two atoms in a definite arrangement held together by very strong (covalent) chemical bonds. Molecule s are distinguished from polyatomic ions in this strict sense. In organic chemistry and biochemistry, the term Molecule is used less strictly and also is applied to charged organic Molecule s and bio Molecule s.
Nuclear force	The nuclear force is the force between two or more nucleons. It is responsible for binding of protons and neutrons into atomic nuclei. To a large extent, this force can be understood in terms of the exchange of virtual light mesons, such as the pions.

Strong

In astronomy in particular, for the description of celestial sources of radiation, Strong is understood to refer to intensity when observed from the Earth or another particular body, relative to other sources relevant in the same observational context. In the optical (visual) range, 'brightness' is synonymous, outside it 'Strong' is standard. Absolute strength, taking into account distance effects, and generally integrating over frequency and over all directions, is termed luminosity, and measured in units of energy per unit time.

Strong interaction

In particle physics, the strong interaction, or strong force, or color force, holds quarks and gluons together to form protons and neutrons. The strong interaction is one of the four fundamental interactions, along with gravitation, the electromagnetic force and the weak interaction. The word strong is used since the strong interaction is the most powerful of the four fundamental forces; its typical field strength is 100 times the strength of the electromagnetic force, some 10^{13} times as great as that of the weak force, and about 10^{38} times that of gravitation.

Weak force

The weak force is one of the four fundamental interactions of nature. In the Standard Model of particle physics, it is due to the exchange of the heavy W and Z bosons. Its most familiar effect is beta decay and the associated radioactivity.

Collision

A Collision is an isolated event in which two or more moving bodies (colliding bodies) exert relatively strong forces on each other for a relatively short time. Deflection happens when an object hits a plane surface

Collision s involve forces (there is a change in velocity.) Collision s can be elastic, meaning they conserve energy and momentum, inelastic, meaning they conserve momentum but not energy, or totally inelastic (or plastic), meaning they conserve momentum and the two objects stick together.

Electromagnetic force

In physics, the Electromagnetic force is the force that the electromagnetic field exerts on electrically charged particles. It is the Electromagnetic force that holds electrons and protons together in atoms, and which hold atoms together to make molecules. The Electromagnetic force operates via the exchange of messenger particles called photons and virtual photons.

Robert Boyle

Robert Boyle (25 January 1627 - 30 December 1691) was an Irish theologian, natural philosopher, chemist, physicist, inventor, and early gentleman scientist, noted for his work in physics and chemistry. He is best known for the formulation of Boyle's law. Although his research and personal philosophy clearly has its roots in the alchemical tradition, he is largely regarded today as the first modern chemist, and therefore one of the founders of modern chemistry.

Robert Hooke

Robert Hooke, FRS was an English natural philosopher and polymath who played an important role in the scientific revolution, through both experimental and theoretical work.

Hooke is known principally for his law of elasticity. He is also remembered for his work as 'the father of microscopy' -- it was Hooke who coined the term 'cell' to describe the basic unit of life -- he also assisted Robert Boyle and built the vacuum pumps used in Boyle's gas law experiments.

Tide

Tide s are the rising of Earth's ocean surface caused by the tidal forces of the Moon and the Sun acting on the oceans. Tide s cause changes in the depth of the marine and estuarine water bodies and produce oscillating currents known as tidal streams, making prediction of Tide s important for coastal navigation . The strip of seashore that is submerged at high Tide and exposed at low Tide the intertidal zone, is an important ecological product of ocean Tide s

Spring tide	A spring tide is when the moon is closest to the Earth during the spring tide. The moon's orbit around the Earth is elliptical which causes the moon to be closer to the Earth and farther away at different times.
Telescope	A telescope is an instrument designed for the observation of remote objects by the collection of electromagnetic radiation. The first known practically functioning telescope s were invented in the Netherlands at the beginning of the 17th century. ' telescope s' can refer to a whole range of instruments operating in most regions of the electromagnetic spectrum.
Reflecting Telescope	A Reflecting telescope is an optical telescope which uses a single or combination of curved mirrors that reflect light and form an image. The Reflecting telescope was invented in the 17th century as an alternative to the refracting telescope which, at that time, was a design that suffered from severe chromatic aberration. Although Reflecting telescope s produce other types of optical aberrations, it is a design that allows for very large diameter objectives.

Big Bang

The Big Bang is the cosmological model of the initial conditions and subsequent development of the universe that is supported by the most comprehensive and accurate explanations from current scientific evidence and observation. As used by cosmologists, the term Big Bang generally refers to the idea that the universe has expanded from a primordial hot and dense initial condition at some finite time in the past (currently estimated to have been approximately 13.7 billion years ago), and continues to expand to this day.

Georges Lemaître proposed what became known as the Big Bang theory of the origin of the Universe, although he called it his 'hypothesis of the primeval atom'.

Conservation law

In physics, a Conservation law states that a particular measurable property of an isolated physical system does not change as the system evolves.

One particularly important physical result concerning Conservation law s is Noether's Theorem, which states that there is a one-to-one correspondence between Conservation law s and differentiable symmetries of physical systems. For example, the conservation of energy follows from the time-invariance of physical systems, and the fact that physical systems behave the same regardless of how they are oriented in space gives rise to the conservation of angular momentum.

Energy

In physics, Energy is a scalar physical quantity that describes the amount of work that can be performed by a force, an attribute of objects and systems that is subject to a conservation law. Different forms of Energy include kinetic, potential, thermal, gravitational, sound, light, elastic, and electromagnetic Energy The forms of Energy are often named after a related force.

Matter

The term Matter traditionally refers to the substance that objects are made of. One common way to identify this 'substance' is through its physical properties; a common definition of Matter is anything that has mass and occupies a volume. However, this definition has to be revised in light of quantum mechanics, where the concept of 'having mass', and 'occupying space' are not as well-defined as in everyday life.

Universe

The Universe is defined as everything that physically exists: the entirety of space and time, all forms of matter, energy and momentum, and the physical laws and constants that govern them. However, the term 'universe' may be used in slightly different contextual senses, denoting such concepts as the cosmos, the world or Nature.

Astronomical observations indicate that the universe is 13.73 ± 0.12 billion years old and at least 93 billion light years across.

Mass

Mass is a concept used in the physical sciences to explain a number of observable behaviours, and in everyday usage, it is common to identify Mass with those resulting behaviors. In particular, Mass is commonly identified with weight. But according to our modern scientific understanding, the weight of an object results from the interaction of its Mass with a gravitational field, so while Mass is part of the explanation of weight, it is not the complete explanation.

Mechanics

Mechanics is the branch of physics concerned with the behaviour of physical bodies when subjected to forces or displacements, and the subsequent effect of the bodies on their environment. The discipline has its roots in several ancient civilizations During the early modern period, scientists such as Galileo, Kepler, and especially Newton, laid the foundation for what is now known as classical Mechanics

Universal gravitation

Newton's law of universal gravitation is a physical law describing the gravitational attraction between bodies with mass. It is a part of classical mechanics and was first formulated in Newton's work Philosophiae Naturalis Principia Mathematica, first published on July 5, 1687. In modern language it states the following:

Every point mass attracts every other point mass by a force pointing along the line intersecting both points.

Isolated System

In the natural sciences an Isolated system as contrasted with a open system, is a physical system that does not interact with its surroundings. It obeys a number of conservation laws: its total energy and mass stay constant. They cannot enter or exit, but can only move around inside.

Second

The second (SI symbol: s), sometimes abbreviated sec., is the name of a unit of time, and is the International System of Units (SI) base unit of time. It may be measured using a clock.

SI prefixes are frequently combined with the word second to denote subdivisions of the second e.g., the milli second (one thousandth of a second , the micro second (one millionth of a second , and the nano second (one billionth of a second)

Linear

The word Linear comes from the Latin word Linear is, which means created by lines. In mathematics, a Linear map or function f(x) is a function which satisfies the following two properties...

· Additivity (also called the superposition property): f(x + y) = f(x) + f(y.)

Vector

In elementary mathematics, physics, and engineering, a vector is a geometric object that has both a magnitude and a direction. A vector is frequently represented by a line segment with a definite direction, or graphically as an arrow, connecting an initial point A with a terminal point B, and denoted by

$$\overrightarrow{AB}.$$

The magnitude of the vector is the length of the segment and the direction characterizes the displacement of B relative to A: how much one should move the point A to 'carry' it to the point B.

Many algebraic operations on real numbers have close analogues for vectors.

Momentum

In classical mechanics, Momentum (pl. momenta; SI unit kgÂ·m/s, or, equivalently, NÂ·s) is the product of the mass and velocity of an object (p = mv.) For more accurate measures of Momentum see the section 'modern definitions of Momentum on this page.

Impulse

In classical mechanics, an Impulse is defined as the integral of a force with respect to time. When a force is applied to a rigid body it changes the momentum of that body. A small force applied for a long time can produce the same momentum change as a large force applied briefly, because it is the product of the force and the time for which it is applied that is important.

Collision

A Collision is an isolated event in which two or more moving bodies (colliding bodies) exert relatively strong forces on each other for a relatively short time. Deflection happens when an object hits a plane surface

Collision s involve forces (there is a change in velocity.) Collision s can be elastic, meaning they conserve energy and momentum, inelastic, meaning they conserve momentum but not energy, or totally inelastic (or plastic), meaning they conserve momentum and the two objects stick together.

Force

In physics, a Force is any external agent that causes a change in the motion of a free body, or that causes stress in a fixed body. It can also be described by intuitive concepts such as a push or pull that can cause an object with mass to change its velocity , i.e., to accelerate, or which can cause a flexible object to deform. Force has both magnitude and direction, making it a vector quantity.

Internal energy

In thermodynamics, the Internal energy of a thermodynamic system denoted by U is the total of the kinetic energy due to the motion of molecules (translational, rotational, vibrational) and the potential energy associated with the vibrational and electric energy of atoms within molecules or crystals. It includes the energy in all of the chemical bonds, and the energy of the free, conduction electrons in metals.

One can also calculate the Internal energy of electromagnetic or blackbody radiation.

Kinetic energy

The Kinetic energy of an object is the extra energy which it possesses due to its motion. It is defined as the work needed to accelerate a body of a given mass from rest to its current velocity. Having gained this energy during its acceleration, the body maintains this Kinetic energy unless its speed changes.

Potential energy

Potential energy can be thought of as energy stored within a physical system. It is called potential energy because it has the potential to be converted into other forms of energy, such as kinetic energy, and to do work in the process. The standard unit of measure for potential energy is the joule, the same as for work, or energy in general.

Work

In physics, mechanical Work is the amount of energy transferred by a force acting through a distance. Like energy, it is a scalar quantity, with SI units of joules. The term Work was first coined in the 1830s by the French mathematician Gaspard-Gustave Coriolis.

Centripetal force

Centripetal force is a force that makes a body follow a curved, as opposed to straight, path; it is always directed orthogonal to the velocity of the body, toward the instantaneous center of curvature of the path. The term Centripetal force comes from the Latin words centrum and petere ('tend towards', 'aim at'), signifying that the force is directed inward toward the center of curvature of the path. Isaac Newton's description was: 'A Centripetal force is that by which bodies are drawn or impelled, or in any way tend, towards a point as to a center.' and is found in the Principia.

Mechanical energy

In physics, Mechanical energy describes the sum of potential energy and kinetic energy present in the components of a mechanical system.

Scientists make simplifying assumptions to make calculations about how mechanical systems react. For example, instead of calculating the Mechanical energy separately for each of the billions of molecules in a soccer ball, it is easier to treat the entire ball as one object.

Acceleration

In physics, and more specifically kinematics, Acceleration is the change in velocity over time. Because velocity is a vector, it can change in two ways: a change in magnitude and/or a change in direction. In one dimension, Acceleration is the rate at which something speeds up or slows down.

Gravitational potential

In celestial mechanics, the Gravitational potential is a scalar field created by any mass, such as the earth or the Sun. The potential at a distance r from a point mass M is given by

$$P = -\frac{GM}{r}$$

where G is the gravitational constant.

The gravitational field is the gradient of this potential, and follows an inverse square law.

Air

The Earth's atmosphere is a layer of gases surrounding the planet Earth that is retained by Earth's gravity. The atmosphere protects life on Earth by absorbing ultraviolet solar radiation, warming the surface through heat retention (greenhouse effect), and reducing temperature extremes between day and night. Dry Air contains roughly (by volume) 78.08% nitrogen, 20.95% oxygen, 0.93% argon, 0.038% carbon dioxide, and trace amounts of other gases.

Gas

In physics, a Gas is a state of matter, consisting of a collection of particles (molecules, atoms, ions, electrons, etc.) without a definite shape or volume that are in more or less random motion. Gas ses are generally modeled by the ideal Gas law where the total volume of the Gas increases proportionally to absolute temperature and decreases inversely proportionally to pressure.

Liquid

Liquid is one of the principal states of matter. A Liquid is a fluid that has the particles loose and can freely form a distinct surface at the boundaries of its bulk material. The surface is a free surface where the Liquid is not constrained by a container.

Molecule

A Molecule is defined as a sufficiently stable, electrically neutral group of at least two atoms in a definite arrangement held together by very strong (covalent) chemical bonds. Molecule s are distinguished from polyatomic ions in this strict sense. In organic chemistry and biochemistry, the term Molecule is used less strictly and also is applied to charged organic Molecule s and bio Molecule s.

Temperature

Temperature is a physical property of a system that underlies the common notions of hot and cold; something that is hotter generally has the greater temperature. Specifically, temperature is a property of matter. Temperature is one of the principal parameters of thermodynamics.

Nuclear energy

Nuclear Energy is released by the splitting or merging together of the nuclei of ato. The conversion of nuclear mass to energy is consistent with the mass-energy equivalence formula $\Delta E = \Delta m.c^2$, in which ΔE = energy release, Δm = mass defect, and c = the speed of light in a vacuum. Nuclear energy was first discovered by French physicist Henri Becquerel in 1896, when he found that photographic plates stored in the dark near uranium were blackened like X-ray plates, which had been just recently discovered at the time 1895.

Wind generator

A Wind generator is a device that generates electrical power from wind energy.

Wind generator s have traditionally been wind turbines, i.e. a propeller attached to an electric generator attached to appropriate electronics to attach it to the electrical grid or to charge batteries.

Recently, however, a reciprocating non-turbine Wind generator the Windbelt, has been invented.

Pendulum

A pendulum is a mass that is attached to a pivot, from which it can swing freely. This object is subject to a restoring force due to gravity that will accelerate it toward an equilibrium position. When the pendulum is displaced from its place of rest, the restoring force will cause the pendulum to oscillate about the equilibrium position.

Specular reflection

Specular reflection is the perfect, mirror-like reflection of light from a surface, in which light from a single incoming direction is reflected into a single outgoing direction. Such behavior is described by the law of reflection, which states that the direction of incoming light, and the direction of outgoing light reflected make the same angle with respect to the surface normal, thus the angle of incidence equals the angle of reflection; this is commonly stated as $>\theta_i = >\theta_r$. This behavior was first discovered through careful observation and measurement by Hero of Alexandria.>

This is in contrast to diffuse reflection, where incoming light is reflected in a broad range of directions.

Electron

An Electron is a subatomic particle that carries a negative electric charge. It has no known substructure and is believed to be a point particle. An Electron has a mass that is approximately 1836 times less than that of the proton.

Escape velocity

In physics, Escape velocity is the speed where the kinetic energy of an object is equal to the magnitude of its gravitational potential energy, as calculated by the equation,

$$U_g = \frac{-Gm_1 m_2}{r}$$

It is commonly described as the speed needed to 'break free' from a gravitational field (without any additional impulse) and is theoretical, totally neglecting atmospheric friction. The term Escape velocity can be considered a misnomer because it is actually a speed rather than a velocity, i.e. it specifies how fast the object must move but the direction of movement is irrelevant, unless 'downward.' In more technical terms, Escape velocity is a scalar (and not a vector.)

Lightning

Lightning is an atmospheric discharge of electricity accompanied by thunder, which typically occurs during thunderstorms, and sometimes during volcanic eruptions or dust storms. In the atmospheric electrical discharge, a leader of a bolt of Lightning can travel at speeds of 60,000 m/s (130,000 mph), and can reach temperatures approaching 30,000 °C (54,000 °F), hot enough to fuse silica sand into glass channels known as fulgurites which are normally hollow and can extend some distance into the ground. There are some 16 million Lightning storms in the world every year.

Spark

In mathematics, specifically in linear algebra, the Spark of a matrix A is the smallest number n such that there exists a set of n columns in A which are linearly dependent. Formally,

$$\mathrm{spark}(A) = \min_{d \neq 0} \|d\|_0 \text{ s.t. } Ad = 0.$$

By contrast, the rank of a matrix is the smallest number k such that all sets of k + 1 columns in A are linearly dependent. The concept of the Spark is of use in the theory of compressive sensing, where requirements on the Spark of the measurement matrix are used to ensure stability and consistency of various estimation techniques.

Velocity

In physics, velocity is defined as the rate of change of position. It is a vector physical quantity; both speed and direction are required to define it. In the SI system, it is measured in meters per second: or ms^{-1}.

Orbit

In physics, an orbit is the gravitationally curved path of one object around a point or another body, for example the gravitational orbit of a planet around a star.

Historically, the apparent motion of the planets were first understood in terms of epicycles, which are the sums of numerous circular motions. This predicted the path of the planets quite well, until Johannes Kepler was able to show that the motion of the planets were in fact elliptical motions.

Orbits

The velocity relationship of two objects with mass can thus be considered in four practical classes, with subtypes:

- No orbit
- Interrupted orbits

- Range of interrupted elliptical paths
- Circumnavigating orbits

- Range of elliptical paths with closest point opposite firing point
- Circular path
- Range of elliptical paths with closest point at firing point
- Infinite orbits

- Parabolic paths
- Hyperbolic paths

In many situations relativistic effects can be neglected, and Newton's laws give a highly accurate description of the motion. Then the acceleration of each body is equal to the sum of the gravitational forces on it, divided by its mass, and the gravitational force between each pair of bodies is proportional to the product of their masses and decreases inversely with the square of the distance between them. To this Newtonian approximation, for a system of two point masses or spherical bodies, only influenced by their mutual gravitation, the orbits can be exactly calculated. If the heavier body is much more massive than the smaller, as for a satellite or small moon orbiting a planet or for the Earth orbiting the Sun, it is accurate and convenient to describe the motion in a coordinate system that is centered on the heavier body, and we can say that the lighter body is in orbit around the heavier.

Inelastic collision

An Inelastic collision is a collision in which kinetic energy is not conserved

In collisions of macroscopic bodies, some kinetic energy is turned into vibrational energy of the atoms, causing a heating effect, and the bodies are deformed.

The molecules of a gas or liquid rarely experience perfectly elastic collisions because kinetic energy is exchanged between the molecules' translational motion and their internal degrees of freedom with each collision.

Nucleus

The nucleus of an atom is the very dense region, consisting of nucleons, at the center of an atom. Although the size of the nucleus varies considerably according to the mass of the atom, the size of the entire atom is comparatively constant. Almost all of the mass in an atom is made up from the protons and neutrons in the nucleus with a very small contribution from the orbiting electrons.

Oxygen

Oxygen and -γενÎ®ς (-genÄ"s) (producer, literally begetter) is the element with atomic number 8 and represented by the symbol O. It is a member of the chalcogen group on the periodic table, and is a highly reactive nonmetallic period 2 element that readily forms compounds (notably oxides) with almost all other elements. At standard temperature and pressure two atoms of the element bind to form di Oxygen , a colorless, odorless, tasteless diatomic gas with the formula O_2. Oxygen is the third most abundant element in the universe by mass after hydrogen and helium and the most abundant element by mass in the Earth's crust.

Subatomic

A subatomic particle is an elementary or composite particle smaller than an atom. Particle physics and nuclear physics are concerned with the study of these particles, their interactions, and non-atomic matter.

subatomic particles include the atomic constituents electrons, protons, and neutrons.

Subatomic particle

A subatomic particle is an elementary or composite particle smaller than an atom. Particle physics and nuclear physics are concerned with the study of these particles, their interactions, and non-atomic matter.

Subatomic particles include the atomic constituents electrons, protons, and neutrons.

Subatomic particles

In physics, subatomic particles are the particles composing nucleons and atoms. There are two types of subatomic particles elementary particles, which are not made of other particles, and composite particles. Particle physics and nuclear physics study these particles and how they interact.

Gravity assist

In orbital mechanics and aerospace engineering, a gravitational slingshot, Gravity assist or swing-by is the use of the relative movement and gravity of a planet or other celestial body to alter the path and speed of a spacecraft, typically in order to save fuel, time, and expense. Gravity assist can be used to decelerate or accelerate a spacecraft. The 'assist' is provided by the motion (orbital angular momentum) of the planet as it pulls on the spacecraft.

Solar

Solar power uses Solar Radiation emitted from our sun. Solar power, a renewable energy source, has been used in many traditional technologies for centuries, and is in widespread use where other power supplies are absent, such as in remote locations and in space.

Solar system

The Solar System consists of the Sun and those celestial objects bound to it by gravity. These objects are the eight planets, their 166 known moons, five dwarf planets, and billions of small bodies. The small bodies include asteroids, icy Kuiper belt objects, comets, meteoroids, and interplanetary dust.

Spacecraft

A spacecraft is a vehicle or machine designed for spaceflight. On a sub-orbital spaceflight, a spacecraft enters outer space but then returns to the planetary surface without making a complete orbit. For an orbital spaceflight, a spacecraft enters a closed orbit around the planetary body.

Heat

In physics and thermodynamics, Heat is the process of energy transfer from one body or system due to thermal contact, which in turn is defined as an energy transfer to a body in any other way than due to work performed on the body.

When an infinitesimal amount of Heat δQ is tranferred to a body in thermal equilibrium at absolute temperature T in a reversible way, then it is given by the quantity TdS, where S is the entropy of the body.

A related term is thermal energy, loosely defined as the energy of a body that increases with its temperature.

Watt	The Watt (symbol: W) is a derived unit of power in the International System of Units (SI.) It measures rate of energy conversion. One Watt is equivalent to 1 joule (J) of energy per second.
Angular momentum	Angular momentum is a quantity that is useful in describing the rotational state of a physical system. For a rigid body rotating around an axis of symmetry (e.g. the fins of a ceiling fan), the Angular momentum can be expressed as the product of the body's moment of inertia and its angular velocity ($\mathbf{L} = I\boldsymbol{\omega}$.) In this way, Angular momentum is sometimes described as the rotational analog of linear momentum.
Torque	A torque in physics is a pseudo-vector that measures the tendency of a force to rotate an object about some axis. The magnitude of a torque is defined as the product of a force and the length of the lever arm. Just as a force is a push or a pull, a torque can be thought of as a twist.
Pulsars	Pulsars are highly magnetized rotating neutron stars that emit a beam of electromagnetic radiation in the form of radio waves. Their observed periods range from 1.4 ms to 8.5 s. The radiation can only be observed when the beam of emission is pointing towards the Earth.
Radio	Radio is the transmission of signals, by modulation of electromagnetic waves with frequencies below those of visible light. Electromagnetic radiation travels by means of oscillating electromagnetic fields that pass through the air and the vacuum of space. Information is carried by systematically changing some property of the radiated waves, such as amplitude, frequency, or phase.
Radio waves	Radio waves are electromagnetic waves occurring on the radio frequency portion of the electromagnetic spectrum. A common use is to transport information through the atmosphere or outer space without wires. Radio waves are distinguished from other kinds of electromagnetic waves by their wavelength, a relatively long wavelength in the electromagnetic spectrum.
Star	A star is a massive, luminous ball of plasma that is held together by gravity. The nearest star to Earth is the Sun, which is the source of most of the energy on Earth. Other star s are visible in the night sky, when they are not outshone by the Sun.
Starquake	A starquake is an astrophysical phenomenon that occurs when the crust of a neutron star undergoes a sudden adjustment, analogous to an earthquake on Earth.
Supernova	A supernova is a stellar explosion. They are extremely luminous and cause a burst of radiation that often briefly outshines an entire galaxy before fading from view over several weeks or months. During this short interval, a supernova can radiate as much energy as the Sun could emit over its life span.
Wave	A Wave is a disturbance that propagates through space and time, usually with transference of energy. A mechanical Wave is a Wave that propagates or travels through a medium due to the restoring forces it produces upon deformation. There also exist Wave s capable of traveling through a vacuum, including electromagnetic radiation and probably gravitational radiation.

Waves	Waves is a three-part novel by Ogan Gurel published in 2009. A 21st century version of Faust, the novel explores good and evil in both individual and global settings focusing around a hypothetical technology that has both medical and military applications. The protagonist, Tomas Twarok, is a contemplative and idealistic doctor-turned-entrepreneur who strikes a deal with his college friend, Maximilian Iblis, a ruthless hedge fund manager.
Hot spots	Hot spots in subatomic physics are regions of high energy density or temperature in hadronic or nuclear matter.
Astronomy	Astronomy , 'law') is the scientific study of celestial objects (such as stars, planets, comets, and galaxies) and phenomena that originate outside the Earth's atmosphere (such as the cosmic background radiation.) It is concerned with the evolution, physics, chemistry, meteorology, and motion of celestial objects, as well as the formation and development of the universe.
Optics	Optics is the science that describes the behavior and properties of light and the interaction of light with matter. Optics explains optical phenomena. The word optics comes from á½€πτικÎ®, meaning appearance or look in Ancient Greek.)

Matter

The term Matter traditionally refers to the substance that objects are made of. One common way to identify this 'substance' is through its physical properties; a common definition of Matter is anything that has mass and occupies a volume. However, this definition has to be revised in light of quantum mechanics, where the concept of 'having mass', and 'occupying space' are not as well-defined as in everyday life.

Gas

In physics, a Gas is a state of matter, consisting of a collection of particles (molecules, atoms, ions, electrons, etc.) without a definite shape or volume that are in more or less random motion. Gas ses are generally modeled by the ideal Gas law where the total volume of the Gas increases proportionally to absolute temperature and decreases inversely proportionally to pressure.

Liquid

Liquid is one of the principal states of matter. A Liquid is a fluid that has the particles loose and can freely form a distinct surface at the boundaries of its bulk material. The surface is a free surface where the Liquid is not constrained by a container.

Nuclear fusion

In physics and nuclear chemistry, nuclear fusion is the process by which multiple like-charged atomic nuclei join together to form a heavier nucleus. It is accompanied by the release or absorption of energy. Iron and nickel nuclei have the largest binding energies per nucleon of all nuclei.

Phase

A phase is one part or portion in recurring or serial activities or occurrences logically connected within a greater process, often resulting in an output or a change. Phase or phases may also refer to:

· Phase, a physically distinctive form of a substance, such as the solid, liquid, and gaseous states of ordinary matter
· Phase transition is the transformation of a thermodynamic system from one phase to another
· The initial condition of a cyclic phenomenon

· Phase, initial angle of a sinusoid function at its origin
· Continuous Fourier transform, angle of a complex coefficient representing the phase of one sinusoidal component
· The current state of a cyclic phenomenon

· Lunar phase, the appearance of the Moon as viewed from the Earth
· Planetary phase, the appearance of the illuminated section of a planet
· Instantaneous phase, generalization for both cyclic and non-cyclic phenomena
· Phase factor, a complex scalar in the context of quantum mechanics
· Polyphase system, a means of distributing alternating current electric power in multiple conducting wires with definite phase offsets

· Single-phase electric power
· Three-phase electric power Basics of three-phase electric power
· Three-phase Mathematics of three-phase electric power
· In biology, a part of the cell cycle in which cells divide and reproduce
· Phaser, an audio effect.
· Archaeological phase, a discrete period of occupation at an archaeological site.

· 'Phases', an episode of the TV series Buffy the Vampire Slayer
· Phases, fictional boss monsters from the .hack franchise
· Phase, an incarnation of the DC Comics character usually known as Phantom Girl
· Phase IV, a 1974 science fiction movie directed by Saul Bass

· A phase is a musical composition using Steve Reich's phasing technique.
· Phase, a syntactic domain hypothesized by Noam Chomsky
· Phase 10, a card game created by Fundex Games
· Phase, a music game for the iPod developed by Harmonix Music Systems
· Phase usually a period of combat within a larger military operation .

Plasma

In physics and chemistry, plasma is an ionized gas, in which a certain proportion of electrons are free rather than being bound to an atom or molecule. The ability of the positive and negative charges to move somewhat independently makes the plasma electrically conductive so that it responds strongly to electromagnetic fields. Plasma therefore has properties quite unlike those of solids, liquids or gases and is considered to be a distinct state of matter.

Solid

The solid state of matter is characterized by a distinct structural rigidity and virtual resistance to deformation (i.e. changes of shape and/or volume.) Most solid s have high values both of Young's modulus and of the shear modulus of elasticity. This contrasts with most liquids or fluids, which have a low shear modulus, and typically exhibit the capacity for macroscopic viscous flow.

Star

A star is a massive, luminous ball of plasma that is held together by gravity. The nearest star to Earth is the Sun, which is the source of most of the energy on Earth. Other star s are visible in the night sky, when they are not outshone by the Sun.

Air

The Earth's atmosphere is a layer of gases surrounding the planet Earth that is retained by Earth's gravity. The atmosphere protects life on Earth by absorbing ultraviolet solar radiation, warming the surface through heat retention (greenhouse effect), and reducing temperature extremes between day and night. Dry Air contains roughly (by volume) 78.08% nitrogen, 20.95% oxygen, 0.93% argon, 0.038% carbon dioxide, and trace amounts of other gases.

Atom

The Atom is a basic unit of matter consisting of a dense, central nucleus surrounded by a cloud of negatively charged electrons. The Atom ic nucleus contains a mix of positively charged protons and electrically neutral neutrons (except in the case of hydrogen-1, which is the only stable nuclide with no neutron.) The electrons of an Atom are bound to the nucleus by the electromagnetic force.

Centripetal force

Centripetal force is a force that makes a body follow a curved, as opposed to straight, path; it is always directed orthogonal to the velocity of the body, toward the instantaneous center of curvature of the path. The term Centripetal force comes from the Latin words centrum and petere ('tend towards', 'aim at'), signifying that the force is directed inward toward the center of curvature of the path. Isaac Newton's description was: 'A Centripetal force is that by which bodies are drawn or impelled, or in any way tend, towards a point as to a center.' and is found in the Principia.

Electron

An Electron is a subatomic particle that carries a negative electric charge. It has no known substructure and is believed to be a point particle. An Electron has a mass that is approximately 1836 times less than that of the proton.

Fluid

A Fluid is defined as a substance that continually deforms (flows) under an applied shear stress. All gases are Fluid s, but not all liquids are Fluid s. Fluid s are a subset of the phases of matter and include liquids, gases, plasmas and, to some extent, plastic solids.

Thermal neutron

A thermal neutron is a free neutron that is Boltzmann distributed with kT = 0.024 eV (4.0×10^{-21} J) at room temperature. This gives characteristic (not average, or median) speed of 2.2 km/s. The name 'thermal' comes from their energy being that of the room temperature gas or material they are permeating.

Nucleus

The nucleus of an atom is the very dense region, consisting of nucleons, at the center of an atom. Although the size of the nucleus varies considerably according to the mass of the atom, the size of the entire atom is comparatively constant. Almost all of the mass in an atom is made up from the protons and neutrons in the nucleus with a very small contribution from the orbiting electrons.

Orbit

In physics, an orbit is the gravitationally curved path of one object around a point or another body, for example the gravitational orbit of a planet around a star.

Historically, the apparent motion of the planets were first understood in terms of epicycles, which are the sums of numerous circular motions. This predicted the path of the planets quite well, until Johannes Kepler was able to show that the motion of the planets were in fact elliptical motions.

Orbits

The velocity relationship of two objects with mass can thus be considered in four practical classes, with subtypes:

- No orbit
- Interrupted orbits

- Range of interrupted elliptical paths
- Circumnavigating orbits

- Range of elliptical paths with closest point opposite firing point
- Circular path
- Range of elliptical paths with closest point at firing point
- Infinite orbits

- Parabolic paths
- Hyperbolic paths

In many situations relativistic effects can be neglected, and Newton's laws give a highly accurate description of the motion. Then the acceleration of each body is equal to the sum of the gravitational forces on it, divided by its mass, and the gravitational force between each pair of bodies is proportional to the product of their masses and decreases inversely with the square of the distance between them. To this Newtonian approximation, for a system of two point masses or spherical bodies, only influenced by their mutual gravitation, the orbits can be exactly calculated. If the heavier body is much more massive than the smaller, as for a satellite or small moon orbiting a planet or for the Earth orbiting the Sun, it is accurate and convenient to describe the motion in a coordinate system that is centered on the heavier body, and we can say that the lighter body is in orbit around the heavier.

Proton

The Proton is a subatomic particle with an electric charge of +1 elementary charge. It is found in the nucleus of each atom but is also stable by itself and has a second identity as the hydrogen ion, $^{1}H^{+}$. It is composed of three even more fundamental particles comprising two up quarks and one down quark.

Refrigerator

A Refrigerator (often called a 'fridge' for short) is a cooling appliance comprising a thermally insulated compartment and a heat pump--chemical or mechanical means--to transfer heat from it to the external environment, cooling the contents to a temperature below ambient. Refrigerator s are extensively used to store foods which spoil from bacterial growth if not refrigerated. A device described as a Refrigerator maintains a temperature a few degrees above the freezing point of water; a similar device which maintains a temperature below the freezing point of water is called a 'freezer.' The Refrigerator is a relatively modern invention among kitchen appliances.

Temperature

Temperature is a physical property of a system that underlies the common notions of hot and cold; something that is hotter generally has the greater temperature. Specifically, temperature is a property of matter. Temperature is one of the principal parameters of thermodynamics.

Symbol

A symbol is something such as an object, picture, written word, sound resemblance a red octagon may stand for 'STOP'. On maps, crossed sabres may indicate a battlefield.

Molecule

A Molecule is defined as a sufficiently stable, electrically neutral group of at least two atoms in a definite arrangement held together by very strong (covalent) chemical bonds. Molecule s are distinguished from polyatomic ions in this strict sense. In organic chemistry and biochemistry, the term Molecule is used less strictly and also is applied to charged organic Molecule s and bio Molecule s.

Oxygen

Oxygen and -γεvÎ®ς (-genÄ"s) (producer, literally begetter) is the element with atomic number 8 and represented by the symbol O. It is a member of the chalcogen group on the periodic table, and is a highly reactive nonmetallic period 2 element that readily forms compounds (notably oxides) with almost all other elements. At standard temperature and pressure two atoms of the element bind to form di Oxygen , a colorless, odorless, tasteless diatomic gas with the formula O_2. Oxygen is the third most abundant element in the universe by mass after hydrogen and helium and the most abundant element by mass in the Earth's crust.

Energy

In physics, Energy is a scalar physical quantity that describes the amount of work that can be performed by a force, an attribute of objects and systems that is subject to a conservation law. Different forms of Energy include kinetic, potential, thermal, gravitational, sound, light, elastic, and electromagnetic Energy The forms of Energy are often named after a related force.

Fire

Fire is the rapid oxidation of a combustible material releasing heat, light, and various reaction products such as carbon dioxide and water. If hot enough, the gases may become ionized to produce plasma. Depending on the substances alight, and any impurities outside, the color of the flame and the Fire s intensity might vary.

Nitrogen

Nitrogen is a chemical element that has the symbol N and atomic number 7 and atomic weight 14.0067. Elemental nitrogen is a colorless, odorless, tasteless and mostly inert diatomic gas at standard conditions, constituting 80% by volume of Earth's atmosphere.

Many industrially important compounds, such as ammonia, nitric acid, organic nitrates, and cyanides, contain nitrogen.

Ozone

Ozone or trioxygen is a triatomic molecule, consisting of three oxygen atoms. It is an allotrope of oxygen that is much less stable than the diatomic O_2. Ground-level ozone is an air pollutant with harmful effects on the respiratory systems of animals and humans.

Ozone layer

The photochemical mechanisms that give rise to the ozone layer were worked out by the British physicist Sidney Chapman in 1930. Ozone in the earth's stratosphere is created by ultraviolet light striking oxygen molecules containing two oxygen atoms, splitting them into individual oxygen atoms; surface. About 90% of the ozone in our atmosphere is contained in the stratosphere.

Salt

Salt is a dietary mineral composed primarily of sodium chloride that is essential for animal life, but toxic to most land plants. Salt flavor is one of the basic tastes, and salt is the most popular food seasoning. Salt is also an important preservative.

Salt for human consumption is produced in different forms: unrefined salt, refined salt, and iodized salt.

Sodium

Sodium is an element which has the symbol N, atomic number 11, atomic mass 22.9898 g/mol, common oxidation number +1. Sodium is a soft, silvery white, highly reactive element and is a member of the alkali metals within 'group 1'. It has only one stable isotope, ^{23}Na.

Pollution

Pollution is the introduction of contaminants into an environment that causes instability, disorder, harm or discomfort to the physical systems or living organisms they are in. Pollution can take the form of chemical substances, or energy, such as noise, heat, or light energy. Pollutants, the elements of pollution, can be foreign substances or energies, or naturally occurring; when naturally occurring, they are considered contaminants when they exceed natural levels.

Amorphous solid

An Amorphous solid is a solid in which there is no long-range order of the positions of the atoms. (Solids in which there is long-range atomic order are called crystalline solids or morphous.) Most classes of solid materials can be found or prepared in an amorphous form.

Crystal

A Crystal or Crystal line solid is a solid material whose constituent atoms, molecules, or ions are arranged in an orderly repeating pattern extending in all three spatial dimensions. The scientific study of Crystal s and Crystal formation is Crystal lography. The process of Crystal formation via mechanisms of Crystal growth is called Crystal lization or solidification.

Static

Statics is the branch of mechanics concerned with the analysis of loads on physical systems in static equilibrium, that is, in a state where the relative positions of subsystems do not vary over time, or where components and structures are at rest. When in static equilibrium, the system is either at rest, or its center of mass moves at constant velocity.

By Newton's second law, this situation implies that the net force and net torque on every body in the system is zero, meaning that for every force bearing upon a member, there must be an equal and opposite force.

Static cling

Static Cling is caused by static electricity, usually due to rubbing as in a clothes dryer. It can be removed by deionizing materials with water, and prevented with fabric softener dryer sheets. Antistatic agents are used to make the surfaces slightly conductive, which reduces or prevents the static charge buildup.

Scanning tunneling microscope

Scanning tunneling microscope is a powerful technique for viewing surfaces at the atomic level. Its development in 1981 earned its inventors, Gerd Binnig and Heinrich Rohrer, the Nobel Prize in Physics in 1986 . STM probes the density of states of a material using tunneling current.

Collision

A Collision is an isolated event in which two or more moving bodies (colliding bodies) exert relatively strong forces on each other for a relatively short time. Deflection happens when an object hits a plane surface

Collision s involve forces (there is a change in velocity.) Collision s can be elastic, meaning they conserve energy and momentum, inelastic, meaning they conserve momentum but not energy, or totally inelastic (or plastic), meaning they conserve momentum and the two objects stick together.

R. Buckminster Fuller

R. Buckminster Fuller was an American visionary, designer, architect, poet, author, and inventor.

Liquid crystals

Liquid crystals (Liquid crystals s) are substances that exhibit a phase of matter that has properties between those of a conventional liquid and those of a solid crystal. For instance, an Liquid crystals may flow like a liquid, but its molecules may be oriented in a crystal-like way. There are many different types of Liquid crystals phases, which can be distinguished by their different optical properties (such as birefringence.)

Periodic table

The periodic table of the chemical elements is a tabular method of displaying the chemical elements. Although precursors to this table exist, its invention is generally credited to Russian chemist Dmitri Mendeleev in 1869. Mendeleev intended the table to illustrate recurring trends in the properties of the elements.

Beta decay

In nuclear physics, Beta decay is a type of radioactive decay in which a beta particle (an electron or a positron) is emitted. In the case of electron emission, it is referred to as beta minus (β^-), while in the case of a positron emission as beta plus (β^+.) Kinetic energy of beta particles has continuous spectrum ranging from 0 to maximal available energy (Q), which depends on parent and daughter nuclear states participating in the decay.

Nuclear fission

Nuclear fission is the splitting of the nucleus of an atom into parts often producing free neutrons and other smaller nuclei, which may eventually produce photons. Fission of heavy elements is an exothermic reaction which can release large amounts of energy both as electromagnetic radiation and as kinetic energy of the fragments. Fission is a form of elemental transmutation because the resulting fragments are not the same element as the original atom.

Quantum

In physics, a quantum is an indivisible entity of a quantity that has the same units as the Planck constant and is related to both energy and momentum of elementary particles of matter and of photons and other bosons. The word comes from the Latin 'quantus,' for 'how much.' Behind this, one finds the fundamental notion that a physical property may be 'quantized', referred to as 'quantization'. This means that the magnitude can take on only certain discrete numerical values, rather than any value, at least within a range.

Quantum mechanics	Quantum mechanics is the study of mechanical systems whose dimensions are close to the atomic scale, such as molecules, atoms, electrons, protons and other subatomic particles. Quantum mechanics is a fundamental branch of physics with wide applications. Quantum theory generalizes classical mechanics to provide accurate descriptions for many previously unexplained phenomena such as black body radiation and stable electron orbits.
Radioactivity	Radioactivity can be used in life sciences as a radiolabel to easily visualise components or target molecules in a biological system. Radionuclei are synthesised in particle accelerators and have short half-lives, giving them high maximum theoretical specific activities. This lowers the detection time compared to radionuclei with longer half-lives, such as carbon-14.
Rutherford	The Rutherford is an obsolete unit of radioactivity, defined as the activity of a quantity of radioactive material in which one million nuclei decay per second. It is therefore equivalent to one megabecquerel. It was named after Ernest Rutherford It is not an SI unit.
Transuranium elements	In chemistry, transuranium elements are the chemical elements with atomic numbers greater than 92. Of the elements with atomic numbers 1 to 92, all but four occur in easily detectable quantities on earth, having stable, or very long half life isotopes, or are created as common products of the decay of uranium. All of the elements with higher atomic numbers, however, have been first discovered artificially, and other than plutonium and neptunium, none occur naturally on earth.
Conservation law	In physics, a Conservation law states that a particular measurable property of an isolated physical system does not change as the system evolves. One particularly important physical result concerning Conservation law s is Noether's Theorem, which states that there is a one-to-one correspondence between Conservation law s and differentiable symmetries of physical systems. For example, the conservation of energy follows from the time-invariance of physical systems, and the fact that physical systems behave the same regardless of how they are oriented in space gives rise to the conservation of angular momentum.
Density	The Density of a material is defined as its mass per unit volume. The symbol of Density is ρ '>rho.) Mathematically: $\rho = \frac{m}{V}$ where: ρ (rho) is the Density m is the mass, V is the volume.
Fluid pressure	Fluid pressure is the pressure at some point within a fluid, such as water or air.

· an open condition, such as the ocean, a swimming pool, or the atmosphere; or
· a closed condition, such as a water line or a gas line.

Pressure in open conditions usually can be approximated as the pressure in 'static' or non-moving conditions (even in the ocean where there are waves and currents), because the motions create only negligible changes in the pressure.

Force

In physics, a Force is any external agent that causes a change in the motion of a free body, or that causes stress in a fixed body. It can also be described by intuitive concepts such as a push or pull that can cause an object with mass to change its velocity , i.e., to accelerate, or which can cause a flexible object to deform. Force has both magnitude and direction, making it a vector quantity.

Mass

Mass is a concept used in the physical sciences to explain a number of observable behaviours, and in everyday usage, it is common to identify Mass with those resulting behaviors. In particular, Mass is commonly identified with weight. But according to our modern scientific understanding, the weight of an object results from the interaction of its Mass with a gravitational field, so while Mass is part of the explanation of weight, it is not the complete explanation.

Mechanics

Mechanics is the branch of physics concerned with the behaviour of physical bodies when subjected to forces or displacements, and the subsequent effect of the bodies on their environment. The discipline has its roots in several ancient civilizations During the early modern period, scientists such as Galileo, Kepler, and especially Newton, laid the foundation for what is now known as classical Mechanics

Atmospheric pressure

Atmospheric pressure is sometimes defined as the force per unit area exerted against a surface by the weight of air above that surface at any given point in the Earth's atmosphere. In most circumstances Atmospheric pressure is closely approximated by the hydrostatic pressure caused by the weight of air above the measurement point. Low pressure areas have less atmospheric mass above their location, whereas high pressure areas have more atmospheric mass above their location.

Scalar

A scalar is a variable that only has magnitude, e.g. a speed of 40 km/h. Compare it with vector, a quantity comprising both magnitude and direction, e.g. a velocity of 40km/h north.

· Scalar, a quantity which is independent of viewpoint, a non-tensor
· Scalar, the same as the notion of scalar in differential geometry
· Scalar, an atomic quantity that can hold only one value at a time .

Universal gravitation

Newton's law of universal gravitation is a physical law describing the gravitational attraction between bodies with mass. It is a part of classical mechanics and was first formulated in Newton's work Philosophiae Naturalis Principia Mathematica, first published on July 5, 1687. In modern language it states the following:

Every point mass attracts every other point mass by a force pointing along the line intersecting both points.

Volume

The Volume of any solid, liquid, plasma, vacuum or theoretical object is how much three-dimensional space it occupies, often quantified numerically. One-dimensional figures (such as lines) and two-dimensional shapes such as square geometry squares are assigned zero Volume in the three-dimensional space. Volume is commonly presented in units such as mililitres or cm^3 (milliliters or cubic centimeters.)

Water mass

An oceanographic Water mass is an identifiable body of water which has physical properties distinct from surrounding water. Properties include temperature, salinity, chemical - isotopic ratios, and other physical quantities.

Common Water mass es in the world ocean are: Antarctic Bottom Water (AABW), North Atlantic Deep Water (NADW), Circumpolar Deep Water (CDW), Antarctic Intermediate Water (AAIW), Subantarctic Mode Water (SAMW), Arctic Intermediate Water (AIW), the central waters of various oceanic basins, and various surface waters.

Acceleration

In physics, and more specifically kinematics, Acceleration is the change in velocity over time. Because velocity is a vector, it can change in two ways: a change in magnitude and/or a change in direction. In one dimension, Acceleration is the rate at which something speeds up or slows down.

Linear

The word Linear comes from the Latin word Linear is, which means created by lines. In mathematics, a Linear map or function f(x) is a function which satisfies the following two properties...

· Additivity (also called the superposition property): $f(x + y) = f(x) + f(y.)$

Linear density

Linear density linear mass density or linear mass is a measure of mass per unit of length, and it is a characteristic of strings or other one-dimensional objects. The SI unit of Linear density is the kilogram per metre (kg/m.) The Linear density μ (sometimes denoted by λ), of an object is defined as:

$$\mu = \frac{\partial m}{\partial x}$$

where m is the mass, and x is a coordinate along the (one dimensional) object.

Measurement

The framework of quantum mechanics requires a careful definition of measurement The issue of measurement lies at the heart of the problem of the interpretation of quantum mechanics, for which there is currently no consensus.

Specific gravity

Specific gravity is defined as the ratio of the density of a given solid or liquid substance to the density of water at a specific temperature and pressure, typically at 4°C and 1 atm , making it a dimensionless quantity. Substances with a specific gravity greater than one are denser than water, and so will sink in it, and those with a specific gravity of less than one are less dense than water, and so will float in it. Specific gravity is a special case of, or in some usages synonymous with, relative density, with the latter term often preferred in modern scientific writing.

Surface	In mathematics, specifically in topology, a surface is a two-dimensional topological manifold. The most familiar examples are those that arise as the boundaries of solid objects in ordinary three-dimensional Euclidean space R^3 -- for example, the surface of a ball or bagel. On the other hand, there are surface s which cannot be embedded in three-dimensional Euclidean space without introducing singularities or intersecting itself -- these are the unorientable surface s.
Spacecraft	A spacecraft is a vehicle or machine designed for spaceflight. On a sub-orbital spaceflight, a spacecraft enters outer space but then returns to the planetary surface without making a complete orbit. For an orbital spaceflight, a spacecraft enters a closed orbit around the planetary body.
Principle	A principle is one of several things: (a) a descriptive comprehensive and fundamental law, doctrine, or assumption; (b) a normative rule or code of conduct, and (c) a law or fact of nature underlying the working of an artificial device. The principle of any effect is the cause that produces it. Depending on the way the cause is understood the basic law governing that cause may acquire some distinction in its expression.
Superfluidity	Superfluidity is a phase of matter or description of heat capacity in which unusual effects are observed when liquids, typically of helium-4 or helium-3, overcome friction by surface interaction when at a stage, known as the 'lambda point' for helium-4, at which the liquid's viscosity becomes zero. Also known as a major facet in the study of quantum hydrodynamics, it was discovered by Pyotr Kapitsa, John F. Allen, and Don Misener in 1937 and has been described through phenomenological and microscopic theories.
Viscosity	Viscosity is a measure of the resistance of a fluid which is being deformed by either shear stress or extensional stress. In general terms it is the resistance of a liquid to flow, or its 'thickness'. Viscosity describes a fluid's internal resistance to flow and may be thought of as a measure of fluid friction.
Superconductivity	Superconductivity is a phenomenon occurring in certain materials generally at very low temperatures, characterized by exactly zero electrical resistance and the exclusion of the interior magnetic field. The electrical resistivity of a metallic conductor decreases gradually as the temperature is lowered. However, in ordinary conductors such as copper and silver, impurities and other defects impose a lower limit.
Radon	Radon is the chemical element that has the symbol Rn and atomic number 86. Radon is a colorless, odorless, naturally occurring, radioactive noble gas that is formed from the decay of radium. It is one of the heaviest substances that are gases under normal conditions and is considered to be a health hazard.
Uranium	Uranium is a silvery-gray metallic chemical element in the actinide series of the periodic table that has the symbol U and atomic number 92. It has 92 protons and 92 electrons, 6 of them valence electrons. It can have between 141 and 146 neutrons, with 146 and 143 in its most common isotopes.
Kinetic energy	The Kinetic energy of an object is the extra energy which it possesses due to its motion. It is defined as the work needed to accelerate a body of a given mass from rest to its current velocity. Having gained this energy during its acceleration, the body maintains this Kinetic energy unless its speed changes.

Potential energy	Potential energy can be thought of as energy stored within a physical system. It is called potential energy because it has the potential to be converted into other forms of energy, such as kinetic energy, and to do work in the process. The standard unit of measure for potential energy is the joule, the same as for work, or energy in general.
Atomic theory	In chemistry and physics, Atomic theory is a theory of the nature of matter, which states that matter is composed of discrete units called atoms, as opposed to the obsolete notion that matter could be divided into any arbitrarily small quantity. It began as a philosophical concept in ancient Greece and India and entered the scientific mainstream in the early 19th century when discoveries in the field of chemistry showed that matter did indeed behave as if it were made up of particles. The word 'atom' was applied to the basic particle that constituted a chemical element, because the chemists of the era believed that these were the fundamental particles of matter.
Robert Boyle	*Robert Boyle* (25 January 1627 - 30 December 1691) was an Irish theologian, natural philosopher, chemist, physicist, inventor, and early gentleman scientist, noted for his work in physics and chemistry. He is best known for the formulation of Boyle's law. Although his research and personal philosophy clearly has its roots in the alchemical tradition, he is largely regarded today as the first modern chemist, and therefore one of the founders of modern chemistry.
Subatomic	A subatomic particle is an elementary or composite particle smaller than an atom. Particle physics and nuclear physics are concerned with the study of these particles, their interactions, and non-atomic matter. subatomic particles include the atomic constituents electrons, protons, and neutrons.
Subatomic particle	A subatomic particle is an elementary or composite particle smaller than an atom. Particle physics and nuclear physics are concerned with the study of these particles, their interactions, and non-atomic matter. Subatomic particles include the atomic constituents electrons, protons, and neutrons.
Subatomic particles	In physics, subatomic particles are the particles composing nucleons and atoms. There are two types of subatomic particles elementary particles, which are not made of other particles, and composite particles. Particle physics and nuclear physics study these particles and how they interact.
Vacuum	A vacuum is a volume of space that is essentially empty of matter, such that its gaseous pressure is much less than atmospheric pressure. The word comes from the Latin term for 'empty,' but in reality, no volume of space can ever be perfectly empty. A perfect vacuum with a gaseous pressure of absolute zero is a philosophical concept that is never observed in practice.

Heat

In physics and thermodynamics, Heat is the process of energy transfer from one body or system due to thermal contact, which in turn is defined as an energy transfer to a body in any other way than due to work performed on the body.

When an infinitesimal amount of Heat δQ is tranferred to a body in thermal equilibrium at absolute temperature T in a reversible way, then it is given by the quantity TdS, where S is the entropy of the body.

A related term is thermal energy, loosely defined as the energy of a body that increases with its temperature.

Heat engine

In engineering and thermodynamics, a Heat engine performs the conversion of heat energy to mechanical work by exploiting the temperature gradient between a hot 'source' and a cold 'sink'. Heat is transferred from the source, through the 'working body' of the engine, to the sink, and in this process some of the heat is converted into work by exploiting the properties of a working substance (usually a gas or liquid.) Figure 1: Heat engine diagram

Heat engine s are often confused with the cycles they attempt to mimic.

Solar

Solar power uses Solar Radiation emitted from our sun. Solar power, a renewable energy source, has been used in many traditional technologies for centuries, and is in widespread use where other power supplies are absent, such as in remote locations and in space.

Solar energy

Solar energy is the light and radiant heat from the Sun that influences Earth's climate and weather and sustains life. Solar power is the rate of solar energy at a point in time; it is sometimes used as a synonym for solar energy or more specifically to refer to electricity generated from solar radiation. Since ancient times solar energy has been harnessed for human use through a range of technologies.

Temperature

Temperature is a physical property of a system that underlies the common notions of hot and cold; something that is hotter generally has the greater temperature. Specifically, temperature is a property of matter. Temperature is one of the principal parameters of thermodynamics.

Internal energy

In thermodynamics, the Internal energy of a thermodynamic system denoted by U is the total of the kinetic energy due to the motion of molecules (translational, rotational, vibrational) and the potential energy associated with the vibrational and electric energy of atoms within molecules or crystals. It includes the energy in all of the chemical bonds, and the energy of the free, conduction electrons in metals.

One can also calculate the Internal energy of electromagnetic or blackbody radiation.

Phase

A phase is one part or portion in recurring or serial activities or occurrences logically connected within a greater process, often resulting in an output or a change. Phase or phases may also refer to:

· Phase, a physically distinctive form of a substance, such as the solid, liquid, and gaseous states of ordinary matter
· Phase transition is the transformation of a thermodynamic system from one phase to another
· The initial condition of a cyclic phenomenon

· Phase, initial angle of a sinusoid function at its origin
· Continuous Fourier transform, angle of a complex coefficient representing the phase of one sinusoidal component
· The current state of a cyclic phenomenon

· Lunar phase, the appearance of the Moon as viewed from the Earth
· Planetary phase, the appearance of the illuminated section of a planet
· Instantaneous phase, generalization for both cyclic and non-cyclic phenomena
· Phase factor, a complex scalar in the context of quantum mechanics
· Polyphase system, a means of distributing alternating current electric power in multiple conducting wires with definite phase offsets

· Single-phase electric power
· Three-phase electric power Basics of three-phase electric power
· Three-phase Mathematics of three-phase electric power
· In biology, a part of the cell cycle in which cells divide and reproduce
· Phaser, an audio effect.
· Archaeological phase, a discrete period of occupation at an archaeological site.

· 'Phases', an episode of the TV series Buffy the Vampire Slayer
· Phases, fictional boss monsters from the .hack franchise
· Phase, an incarnation of the DC Comics character usually known as Phantom Girl
· Phase IV, a 1974 science fiction movie directed by Saul Bass

· A phase is a musical composition using Steve Reich's phasing technique.
· Phase, a syntactic domain hypothesized by Noam Chomsky
· Phase 10, a card game created by Fundex Games
· Phase, a music game for the iPod developed by Harmonix Music Systems
· Phase usually a period of combat within a larger military operation .

Phase transition

In thermodynamics, phase transition or phase change is the transformation of a thermodynamic system from one phase to another. The distinguishing characteristic of a phase transition is an abrupt change in one or more physical properties, in particular the heat capacity, with a small change in a thermodynamic variable such as the temperature.

In the English vernacular, the term is most commonly used to describe transitions between solid, liquid and gaseous states of matter, in rare cases including plasma.

Radiation

Radiation, as in physics, is energy in the form of waves or moving subatomic particles emitted by an atom or other body as it changes from a higher energy state to a lower energy state. Radiation can be classified as ionizing or non-ionizing radiation, depending on its effect on atomic matter. The most common use of the word 'radiation' refers to ionizing radiation.

Thermometer

A Thermometer is a device that measures temperature or temperature gradient using a variety of different principles. A Thermometer has two important elements: the temperature sensor (e.g. the bulb on a mercury Thermometer in which some physical change occurs with temperature, plus some means of converting this physical change into a value (e.g. the scale on a mercury Thermometer) Thermometer s increasingly use electronic means to provide a digital display or input to a computer.

Nuclear reactor

A Nuclear reactor is a device in which nuclear chain reactions are initiated, controlled, and sustained at a steady rate.

The most significant use of Nuclear reactor s is as an energy source for the generation of electrical power and for the power in some ships This is usually accomplished by methods that involve using heat from the nuclear reaction to power steam turbines.

Transition Temperature

Transition temperature is the temperature at which material changes from one crystal to another. There are total seven crystal systems and every material is known to exist in one of them. At any critical value of temperature it can shift to any crystal system from the previously existing one.

Absolute zero

Absolute zero is a temperature marked by a 0 entropy configuration. It is the coldest temperature theoretically possible and cannot be reached by artificial or natural means, because it is impossible to decouple a system fully from the rest of the universe. Temperature is an entropically defined quantity that effectively determines the number of thermodynamically accessible states of a system within an energy range.

Earth

Earth is the third planet from the Sun. It is the fifth largest of the eight planets in the solar system, and the largest of the terrestrial planets (non-gas planets) in the Solar System in terms of diameter, mass and density. It is also referred to as the World, the Blue Planet, and Terra.

Collision

A Collision is an isolated event in which two or more moving bodies (colliding bodies) exert relatively strong forces on each other for a relatively short time. Deflection happens when an object hits a plane surface

Collision s involve forces (there is a change in velocity.) Collision s can be elastic, meaning they conserve energy and momentum, inelastic, meaning they conserve momentum but not energy, or totally inelastic (or plastic), meaning they conserve momentum and the two objects stick together.

Kinetic energy

The Kinetic energy of an object is the extra energy which it possesses due to its motion. It is defined as the work needed to accelerate a body of a given mass from rest to its current velocity. Having gained this energy during its acceleration, the body maintains this Kinetic energy unless its speed changes.

Liquid

Liquid is one of the principal states of matter. A Liquid is a fluid that has the particles loose and can freely form a distinct surface at the boundaries of its bulk material. The surface is a free surface where the Liquid is not constrained by a container.

Matter

The term Matter traditionally refers to the substance that objects are made of. One common way to identify this 'substance' is through its physical properties; a common definition of Matter is anything that has mass and occupies a volume. However, this definition has to be revised in light of quantum mechanics, where the concept of 'having mass', and 'occupying space' are not as well-defined as in everyday life.

Molecule

A Molecule is defined as a sufficiently stable, electrically neutral group of at least two atoms in a definite arrangement held together by very strong (covalent) chemical bonds. Molecule s are distinguished from polyatomic ions in this strict sense. In organic chemistry and biochemistry, the term Molecule is used less strictly and also is applied to charged organic Molecule s and bio Molecule s.

Potential energy

Potential energy can be thought of as energy stored within a physical system. It is called potential energy because it has the potential to be converted into other forms of energy, such as kinetic energy, and to do work in the process. The standard unit of measure for potential energy is the joule, the same as for work, or energy in general.

Solid

The solid state of matter is characterized by a distinct structural rigidity and virtual resistance to deformation (i.e. changes of shape and/or volume.) Most solid s have high values both of Young's modulus and of the shear modulus of elasticity. This contrasts with most liquids or fluids, which have a low shear modulus, and typically exhibit the capacity for macroscopic viscous flow.

Angular momentum

Angular momentum is a quantity that is useful in describing the rotational state of a physical system. For a rigid body rotating around an axis of symmetry (e.g. the fins of a ceiling fan), the Angular momentum can be expressed as the product of the body's moment of inertia and its angular velocity ($\mathbf{L} = I\boldsymbol{\omega}$.) In this way, Angular momentum is sometimes described as the rotational analog of linear momentum.

Escape velocity

In physics, Escape velocity is the speed where the kinetic energy of an object is equal to the magnitude of its gravitational potential energy, as calculated by the equation,

$$U_g = \frac{-Gm_1 m_2}{r}$$

It is commonly described as the speed needed to 'break free' from a gravitational field (without any additional impulse) and is theoretical, totally neglecting atmospheric friction. The term Escape velocity can be considered a misnomer because it is actually a speed rather than a velocity, i.e. it specifies how fast the object must move but the direction of movement is irrelevant, unless 'downward.' In more technical terms, Escape velocity is a scalar (and not a vector.)

Linear

The word Linear comes from the Latin word Linear is, which means created by lines. In mathematics, a Linear map or function f(x) is a function which satisfies the following two properties...

· Additivity (also called the superposition property): f(x + y) = f(x) + f(y.)

Velocity

In physics, velocity is defined as the rate of change of position. It is a vector physical quantity; both speed and direction are required to define it. In the SI system, it is measured in meters per second: or ms^{-1}.

Astronomy

Astronomy , 'law') is the scientific study of celestial objects (such as stars, planets, comets, and galaxies) and phenomena that originate outside the Earth's atmosphere (such as the cosmic background radiation.) It is concerned with the evolution, physics, chemistry, meteorology, and motion of celestial objects, as well as the formation and development of the universe.

Mass

Mass is a concept used in the physical sciences to explain a number of observable behaviours, and in everyday usage, it is common to identify Mass with those resulting behaviors. In particular, Mass is commonly identified with weight. But according to our modern scientific understanding, the weight of an object results from the interaction of its Mass with a gravitational field, so while Mass is part of the explanation of weight, it is not the complete explanation.

Orbit

In physics, an orbit is the gravitationally curved path of one object around a point or another body, for example the gravitational orbit of a planet around a star.
Historically, the apparent motion of the planets were first understood in terms of epicycles, which are the sums of numerous circular motions. This predicted the path of the planets quite well, until Johannes Kepler was able to show that the motion of the planets were in fact elliptical motions.

Orbits

The velocity relationship of two objects with mass can thus be considered in four practical classes, with subtypes:

- No orbit
- Interrupted orbits

- Range of interrupted elliptical paths
- Circumnavigating orbits

- Range of elliptical paths with closest point opposite firing point
- Circular path
- Range of elliptical paths with closest point at firing point
- Infinite orbits

- Parabolic paths
- Hyperbolic paths

In many situations relativistic effects can be neglected, and Newton's laws give a highly accurate description of the motion. Then the acceleration of each body is equal to the sum of the gravitational forces on it, divided by its mass, and the gravitational force between each pair of bodies is proportional to the product of their masses and decreases inversely with the square of the distance between them. To this Newtonian approximation, for a system of two point masses or spherical bodies, only influenced by their mutual gravitation, the orbits can be exactly calculated. If the heavier body is much more massive than the smaller, as for a satellite or small moon orbiting a planet or for the Earth orbiting the Sun, it is accurate and convenient to describe the motion in a coordinate system that is centered on the heavier body, and we can say that the lighter body is in orbit around the heavier.

Solar system

The Solar System consists of the Sun and those celestial objects bound to it by gravity. These objects are the eight planets, their 166 known moons, five dwarf planets, and billions of small bodies. The small bodies include asteroids, icy Kuiper belt objects, comets, meteoroids, and interplanetary dust.

Solar wind

The solar wind is a stream of charged particles--a plasma--that are ejected from the upper atmosphere of the sun. It consists mostly of electrons and protons with energies of about 1 keV. These particles are able to escape the sun's gravity, in part because of the high temperature of the corona, but also because of high kinetic energy that particles gain through a process that is not well-understood at this time.

Star
A star is a massive, luminous ball of plasma that is held together by gravity. The nearest star to Earth is the Sun, which is the source of most of the energy on Earth. Other star s are visible in the night sky, when they are not outshone by the Sun.

Electron
An Electron is a subatomic particle that carries a negative electric charge. It has no known substructure and is believed to be a point particle. An Electron has a mass that is approximately 1836 times less than that of the proton.

Length
Length is the long dimension of any object. The Length of a thing is the distance between its ends, its linear extent as measured from end to end. This may be distinguished from height, which is vertical extent, and width or breadth, which are the distance from side to side, measuring across the object at right angles to the Length

Plasma
In physics and chemistry, plasma is an ionized gas, in which a certain proportion of electrons are free rather than being bound to an atom or molecule. The ability of the positive and negative charges to move somewhat independently makes the plasma electrically conductive so that it responds strongly to electromagnetic fields. Plasma therefore has properties quite unlike those of solids, liquids or gases and is considered to be a distinct state of matter.

Superconductivity
Superconductivity is a phenomenon occurring in certain materials generally at very low temperatures, characterized by exactly zero electrical resistance and the exclusion of the interior magnetic field.

The electrical resistivity of a metallic conductor decreases gradually as the temperature is lowered. However, in ordinary conductors such as copper and silver, impurities and other defects impose a lower limit.

Thermal
A thermal column (or thermal is a column of rising air in the lower altitudes of the Earth's atmosphere. thermal s are created by the uneven heating of the Earth's surface from solar radiation, and an example of convection. The Sun warms the ground, which in turn warms the air directly above it.

Thermal expansion
Thermal Expansion is the tendency of matter to change in volume in response to a change in temperature. When a substance is heated, its constituent particles move around more vigorously and by doing so generally maintain a greater average separation. Materials that contract with an increase in temperature are very uncommon; this effect is limited in size, and only occurs within limited temperature ranges.

Alpha decay
Alpha decay is a type of radioactive decay in which an atomic nucleus emits an alpha particle (two protons and two neutrons bound together into a particle identical to a helium nucleus) and transforms (or 'decays') into an atom with a mass number 4 less and atomic number 2 less. For example:

(The second form is preferred because the first form appears electrically unbalanced. Fundamentally, the recoiling nucleus is very quickly stripped of the two extra electrons which give it an unbalanced charge.

Bimetallic strip
A Bimetallic strip is used to convert a temperature change into mechanical displacement. The strip consists of two strips of different metals which expand at different rates as they are heated, usually steel and copper, or in some cases brass instead of copper. The strips are joined together throughout their length either riveting, brazing or welding.

Density
The Density of a material is defined as its mass per unit volume. The symbol of Density is ρ '>rho.)

Mathematically:

$$\rho = \frac{m}{V}$$

where:

ρ (rho) is the Density
m is the mass,
V is the volume.

Volume

The Volume of any solid, liquid, plasma, vacuum or theoretical object is how much three-dimensional space it occupies, often quantified numerically. One-dimensional figures (such as lines) and two-dimensional shapes such as square geometry squares are assigned zero Volume in the three-dimensional space. Volume is commonly presented in units such as mililitres or cm^3 (milliliters or cubic centimeters.)

Ideal gas

An Ideal gas is a theoretical gas composed of a set of randomly-moving point particles that interact only through elastic collisions. The Ideal gas concept is useful because it obeys the Ideal gas law, a simplified equation of state, and is amenable to analysis under statistical mechanics.

At normal ambient conditions such as standard temperature and pressure, most real gases behave qualitatively like an Ideal gas

Thermodynamic

In physics, Thermodynamic s '>power') is the study of the conversion of energy into work and heat and its relation to macroscopic variables such as temperature,volume and pressure. Its underpinnings, based upon statistical predictions of the collective motion of particles from their microscopic behavior, is the field of statistical Thermodynamic s, a branch of statistical mechanics. Historically, Thermodynamic s developed out of need to increase the efficiency of early steam engines.

Thermodynamics

In physics, thermodynamics is the study of the transformation of energy into different forms and its relation to macroscopic variables such as temperature, pressure, and volume. Its underpinnings, based upon statistical predictions of the collective motion of particles from their microscopic behavior, is the field of statistical thermodynamics, a branch of statistical mechanics. Roughly, heat means 'energy in transit' and dynamics relates to 'movement'; thus, in essence thermodynamics studies the movement of energy and how energy instills movement.

First law of Thermodynamics

The First law of thermodynamics an expression of the principle of conservation of energy, states that energy can be transformed , but cannot be created or destroyed. Alternatively:

The First law of thermodynamics says that energy is conserved in any process involving a thermodynamic system and its surroundings. Frequently it is convenient to focus on changes in so-called internal energy (U) and to regard them as due to a combination of heat (Q) added to the system and work done by the system (W.)

Energy

In physics, Energy is a scalar physical quantity that describes the amount of work that can be performed by a force, an attribute of objects and systems that is subject to a conservation law. Different forms of Energy include kinetic, potential, thermal, gravitational, sound, light, elastic, and electromagnetic Energy The forms of Energy are often named after a related force.

Gravitational potential

In celestial mechanics, the Gravitational potential is a scalar field created by any mass, such as the earth or the Sun. The potential at a distance r from a point mass M is given by

$$P = -\frac{GM}{r}$$

where G is the gravitational constant.

The gravitational field is the gradient of this potential, and follows an inverse square law.

Air

The Earth's atmosphere is a layer of gases surrounding the planet Earth that is retained by Earth's gravity. The atmosphere protects life on Earth by absorbing ultraviolet solar radiation, warming the surface through heat retention (greenhouse effect), and reducing temperature extremes between day and night. Dry Air contains roughly (by volume) 78.08% nitrogen, 20.95% oxygen, 0.93% argon, 0.038% carbon dioxide, and trace amounts of other gases.

Conservation law

In physics, a Conservation law states that a particular measurable property of an isolated physical system does not change as the system evolves.

One particularly important physical result concerning Conservation law s is Noether's Theorem, which states that there is a one-to-one correspondence between Conservation law s and differentiable symmetries of physical systems. For example, the conservation of energy follows from the time-invariance of physical systems, and the fact that physical systems behave the same regardless of how they are oriented in space gives rise to the conservation of angular momentum.

Thermal equilibrium

In thermodynamics, a thermodynamic system is said to be in thermodynamic equilibrium when it is in thermal equilibrium, mechanical equilibrium, and chemical equilibrium. The local state of a system at thermodynamic equilibrium is determined by the values of its intensive parameters, as pressure, temperature, etc. Specifically, thermodynamic equilibrium is characterized by the minimum of a thermodynamic potential, such as the Helmholtz free energy.

Work

In physics, mechanical Work is the amount of energy transferred by a force acting through a distance. Like energy, it is a scalar quantity, with SI units of joules. The term Work was first coined in the 1830s by the French mathematician Gaspard-Gustave Coriolis.

Convection

Convection in the most general terms refers to the movement of molecules within fluids (i.e. liquids, gases and rheids.) Convection is one of the major modes of heat transfer and mass transfer. In fluids, convective heat and mass transfer take place through both diffusion - the random Brownian motion of individual particles in the fluid - and by advection, in which matter or heat is transported by the larger-scale motion of currents in the fluid.

Fluid

A Fluid is defined as a substance that continually deforms (flows) under an applied shear stress. All gases are Fluid s, but not all liquids are Fluid s. Fluid s are a subset of the phases of matter and include liquids, gases, plasmas and, to some extent, plastic solids.

Vacuum

A vacuum is a volume of space that is essentially empty of matter, such that its gaseous pressure is much less than atmospheric pressure. The word comes from the Latin term for 'empty,' but in reality, no volume of space can ever be perfectly empty. A perfect vacuum with a gaseous pressure of absolute zero is a philosophical concept that is never observed in practice.

Atmosphere

An Atmosphere is a layer of gases that may surround a material body of sufficient mass, by the gravity of the body, and are retained for a longer duration if gravity is high and the Atmosphere s temperature is low. Some planets consist mainly of various gases, but only their outer layer is their Atmosphere .

The term stellar Atmosphere describes the outer region of a star, and typically includes the portion starting from the opaque photosphere outwards.

Force

In physics, a Force is any external agent that causes a change in the motion of a free body, or that causes stress in a fixed body. It can also be described by intuitive concepts such as a push or pull that can cause an object with mass to change its velocity , i.e., to accelerate, or which can cause a flexible object to deform. Force has both magnitude and direction, making it a vector quantity.

Meteorology

Meteorology is the interdisciplinary scientific study of the atmosphere that focuses on weather processes and forecasting . Studies in the field stretch back millennia, though significant progress in Meteorology did not occur until the eighteenth century. The nineteenth century saw breakthroughs occur after observing networks developed across several countries.

Ocean

An ocean is a major body of saline water, and a principal component of the hydrosphere. Approximately 71% of the Earth's surface (an area of some 361 million square kilometers) is covered by ocean a continuous body of water that is customarily divided into several principal ocean s and smaller seas. More than half of this area is over 3,000 meters (9,800 ft) deep.

Radio

Radio is the transmission of signals, by modulation of electromagnetic waves with frequencies below those of visible light. Electromagnetic radiation travels by means of oscillating electromagnetic fields that pass through the air and the vacuum of space. Information is carried by systematically changing some property of the radiated waves, such as amplitude, frequency, or phase.

Radio waves

Radio waves are electromagnetic waves occurring on the radio frequency portion of the electromagnetic spectrum. A common use is to transport information through the atmosphere or outer space without wires. Radio waves are distinguished from other kinds of electromagnetic waves by their wavelength, a relatively long wavelength in the electromagnetic spectrum.

Ultraviolet

Ultraviolet light is electromagnetic radiation with a wavelength shorter than that of visible light, but longer than X-rays. It is so named because the spectrum consists of electromagnetic waves with frequencies higher than those that humans identify as the color violet.

UV light is typically found as part of the radiation received by the Earth from the Sun.

Visible light

The visible spectrum is the portion of the electromagnetic spectrum that is visible to the human eye. Electromagnetic radiation in this range of wavelengths is called visible light or simply light. A typical human eye will respond to wavelengths in air from about 380 to 750 nm.

Wave

A Wave is a disturbance that propagates through space and time, usually with transference of energy. A mechanical Wave is a Wave that propagates or travels through a medium due to the restoring forces it produces upon deformation. There also exist Wave s capable of traveling through a vacuum, including electromagnetic radiation and probably gravitational radiation.

Waves

Waves is a three-part novel by Ogan Gurel published in 2009. A 21st century version of Faust, the novel explores good and evil in both individual and global settings focusing around a hypothetical technology that has both medical and military applications. The protagonist, Tomas Twarok, is a contemplative and idealistic doctor-turned-entrepreneur who strikes a deal with his college friend, Maximilian Iblis, a ruthless hedge fund manager.

X-radiation

X-radiation is a form of electromagnetic radiation. X-rays have a wavelength in the range of 10 to 0.01 nanometers, corresponding to frequencies in the range 30 petahertz to 30 exahertz and energies in the range 120 eV to 120 keV. They are longer than gamma rays but shorter than UV rays.

Reflection

Reflection is the change in direction of a wave front at an interface between two different media so that the wave front returns into the medium from which it originated. Common examples include the reflection of light, sound and water waves.

Law of reflection: Angle of incidence = Angle of reflection

Reflections may occur in a number of wave and particle phenomena; these include acoustic, seismic waves in geologic structures, surface waves in bodies of water, and various electromagnetic waves, most usefully from VHF and higher radar frequencies, progressing upward through centimeter to millimeter-wavelength radar and the various light frequencies and (with special 'grazing' mirrors, to X-ray frequencies and beyond to gamma rays.

Specific heat

Specific heat capacity, also known simply as specific heat, is the measure of the heat energy required to increase the temperature of a unit quantity of a substance by a certain temperature interval. The term originated primarily through the work of Scottish physicist Joseph Black who conducted various heat measurements and used the phrase 'capacity for heat.' More heat energy is required to increase the temperature of a substance with high specific heat capacity than one with low specific heat capacity. For instance, eight times the heat energy is required to increase the temperature of an ingot of magnesium as is required for a lead ingot of the same mass.

Specific heat capacity	Specific heat capacity is the measure of the heat energy required to increase the temperature of a unit quantity of a substance by a certain temperature interval. The term originated primarily through the work of Scottish physicist Joseph Black who conducted various heat measurements and used the phrase 'capacity for heat.' More heat energy is required to increase the temperature of a substance with high specific heat capacity than one with low specific heat capacity. For instance, eight times the heat energy is required to increase the temperature of an ingot of magnesium as is required for a lead ingot of the same mass.
Gas	In physics, a Gas is a state of matter, consisting of a collection of particles (molecules, atoms, ions, electrons, etc.) without a definite shape or volume that are in more or less random motion. Gas ses are generally modeled by the ideal Gas law where the total volume of the Gas increases proportionally to absolute temperature and decreases inversely proportionally to pressure.
Mechanical energy	In physics, Mechanical energy describes the sum of potential energy and kinetic energy present in the components of a mechanical system. Scientists make simplifying assumptions to make calculations about how mechanical systems react. For example, instead of calculating the Mechanical energy separately for each of the billions of molecules in a soccer ball, it is easier to treat the entire ball as one object.
Milky Way	The Milky Way is the galaxy in which the Solar System is located. It is a barred spiral galaxy that is part of the Local Group of galaxies. It is one of billions of galaxies in the observable universe.
Nuclear fusion	In physics and nuclear chemistry, nuclear fusion is the process by which multiple like-charged atomic nuclei join together to form a heavier nucleus. It is accompanied by the release or absorption of energy. Iron and nickel nuclei have the largest binding energies per nucleon of all nuclei.
White	White is not a color, the perception which is evoked by light that stimulates all three types of color sensitive cone cells in the human eye in near equal amount and with high brightness compared to the surroundings. Since the impression of white is obtained by three summations of light intensity across the visible spectrum, the number of combinations of light wavelengths that produce the sensation of white is practically infinite. There are a number of different white light sources such as the midday Sun, incandescent lamps, fluorescent lamps and white LEDs.
White dwarf	A White dwarf is a small star composed mostly of electron-degenerate matter. Because a White dwarf s mass is comparable to that of the Sun and its volume is comparable to that of the Earth, it is very dense. Its faint luminosity comes from the emission of stored thermal energy.
Opacity	Opacity is the measure of impenetrability to electromagnetic or other kinds of radiation, especially visible light. In radiative transfer, it describes the absorption and scattering of radiation in a medium, such as a plasma, dielectric, shielding material, glass, etc. An opaque object is neither transparent nor translucent.
Inelastic collision	An Inelastic collision is a collision in which kinetic energy is not conserved In collisions of macroscopic bodies, some kinetic energy is turned into vibrational energy of the atoms, causing a heating effect, and the bodies are deformed.

The molecules of a gas or liquid rarely experience perfectly elastic collisions because kinetic energy is exchanged between the molecules' translational motion and their internal degrees of freedom with each collision.

Satellite

In the context of spaceflight, a satellite is an object which has been placed into orbit by human endeavor. Such objects are sometimes called artificial satellite s to distinguish them from natural satellite s such as the Moon.

The first artificial satellite Sputnik 1, was launched by the Soviet Union in 1957.

Nuclear power

Nuclear power is any nuclear technology designed to extract usable energy from atomic nuclei via controlled nuclear reactions. The most common method today is through nuclear fission, though other methods include nuclear fusion and radioactive decay. All utility-scale reactors heat water to produce steam, which is then converted into mechanical work for the purpose of generating electricity or propulsion.

Pressurized water reactor

Pressurized water reactor are generation II nuclear power reactors that use ordinary water under high pressure as coolant and neutron moderator. The primary coolant loop is kept under high pressure to prevent the water from reaching film boiling, hence the name. PWRs are the most common type of power producing reactor and are widely used all over the world.

Refrigerator

A Refrigerator (often called a 'fridge' for short) is a cooling appliance comprising a thermally insulated compartment and a heat pump--chemical or mechanical means--to transfer heat from it to the external environment, cooling the contents to a temperature below ambient. Refrigerator s are extensively used to store foods which spoil from bacterial growth if not refrigerated. A device described as a Refrigerator maintains a temperature a few degrees above the freezing point of water; a similar device which maintains a temperature below the freezing point of water is called a 'freezer.' The Refrigerator is a relatively modern invention among kitchen appliances.

Evaporation

Evaporation is the slow vaporization of a liquid and the reverse of condensation. A type of phase transition, it is the process by which molecules in a liquid state (e.g. water) spontaneously become gaseous (e.g. water vapor.) Generally, Evaporation can be seen by the gradual disappearance of a liquid from a substance when exposed to a significant volume of gas.

Humidity

Humidity is the amount of water vapour in the air. In daily language the term Humidity is normally taken to mean relative Humidity Relative Humidity is defined as the ratio of the partial pressure of water vapour in a parcel of air to the saturated vapour pressure of water vapour at a prescribed temperature.

Latent heat

The expression Latent heat refers to the amount of energy released or absorbed by a chemical substance during a change of state that occurs without changing its temperature, meaning a phase transition such as the melting of ice or the boiling of water. The term was introduced around 1750 by Joseph Black as derived from the Latin latere, to lie hidden.

Two of the more common forms of Latent heat encountered are Latent heat of fusion (melting) and Latent heat of vaporization (boiling.)

Vapor

A vapor is a substance in the gas phase at a temperature lower than its critical temperature. This means that the vapor can be condensed to a liquid or to a solid by increasing its pressure, without reducing the temperature.

For example, water has a critical temperature of 374>°C which is the highest temperature at which liquid water can exist.

Vaporization

Vaporization of an element or compound is a phase transition from the liquid phase to gas phase. There are two sorts of vaporization: evaporation and boiling.

Evaporation is a phase transition from the liquid phase to gas phase that occurs at temperatures below the boiling temperature at a given pressure.

Water vapor

Water vapor or water vapour, also aqueous vapor, is the gas phase of water. Water vapor is one state of the water cycle within the hydrosphere. Water vapor can be produced from the evaporation of liquid water or from the sublimation of ice.

Relative humidity

Relative humidity is a measurement of the amount of water vapor that exists in a gaseous mixture of air and water.

The relative humidity of an air-water vapor mixture can be estimated if both the temperature and the dew point temperature of the mixture are known. When both T and T_d are expressed in degrees celsius then :

$$RH = \frac{e_p}{e_s} \times 100\%$$

where the partial pressure of water vapor in the mixture is estimated by e_p :

$$e_p = e^{\frac{(17.269 \times T_d)}{(273.3+T_d)}}$$

and the saturated vapor pressure of water at the temperature of the mixture is estimated by e_s :

$$e_s = e^{\frac{(17.269 \times T)}{(273.3+T)}}$$

In practice both T and T_d are readily estimated by using a sling psychrometer and the relative humidity of the atmosphere can be calculated

Often the notion of air holding water vapor is used to describe the concept of relative humidity.

Saturation

Seen in some magnetic materials, Saturation is the state reached when an increase in applied external magnetizing field H cannot increase the magnetization of the material further, so the total magnetic field B levels off. It is a characteristic particularly of ferromagnetic materials, such as iron, nickel, cobalt, manganese and their alloys.

Saturation is most clearly seen in the magnetization curve (also called BH curve or hysteresis curve) of a substance, as a bending to the right of the curve

Power station

A power station is an industrial facility for the generation of electric power.

Power plant is also used to refer to the engine in ships, aircraft and other large vehicles. Some prefer to use the term energy center because it more accurately describes what the plants do, which is the conversion of other forms of energy, like chemical energy, gravitational potential energy or heat energy into electrical energy.

Second	The second (SI symbol: s), sometimes abbreviated sec., is the name of a unit of time, and is the International System of Units (SI) base unit of time. It may be measured using a clock. SI prefixes are frequently combined with the word second to denote subdivisions of the second e.g., the milli second (one thousandth of a second , the micro second (one millionth of a second , and the nano second (one billionth of a second)
Second law of Thermodynamics	The second law of thermodynamics is an expression of the universal law of increasing entropy, stating that the entropy of an isolated system which is not in equilibrium will tend to increase over time, approaching a maximum value at equilibrium. The second law traces its origin to French physicist Sadi Carnot's 1824 paper Reflections on the Motive Power of Fire, which presented the view that motive power is due to the fall of caloric from a hot to cold body. In simple terms, the second law is an expression of the fact that over time, ignoring the effects of self-gravity, differences in temperature, pressure, and density tend to even out in a physical system that is isolated from the outside world.
Electric power	Electric power is defined as the rate at which electrical energy is transferred by an electric circuit. The SI unit of power is the watt. Electrical power is distributed via cables and electricity pylons like these in Brisbane, Australia. When electric current flows in a circuit, it can transfer energy to do mechanical or thermodynamic work.
Efficiency	The efficiency of an entity (a device, component typically denoted by the Greek letter small Eta . $$\text{Efficiency} = \frac{\text{Useful power output}}{\text{Total power input}}$$
Thermal pollution	Thermal pollution is a temperature change in natural bodies of water caused by human influence. The temperature change can be upwards or downwards. In the Northern Hemisphere, a common cause of thermal pollution is the use of water as a coolant, especially in power plants.
Refrigerant	A refrigerant is a compound used in a heat cycle that undergoes a phase change from a gas to a liquid and back. The two main uses of refrigerant s are refrigerators/freezers and air conditioners (cf. coolant.)
Electricity	Electricity is a general term that encompasses a variety of phenomena resulting from the presence and flow of electric charge. These include many easily recognizable phenomena, such as lightning and static Electricity but in addition, less familiar concepts, such as the electromagnetic field and electromagnetic induction. In general usage, the word Electricity is adequate to refer to a number of physical effects.
Entropy	Entropy is an information theoretical concept applied across physics, information theory, mathematics and other branches of science and engineering. When given a system whose exact description is not precisely known, the Entropy is defined as the expected amount of information needed to exactly specify the state of the system, given what we know about the system. The Entropy (S) can be computed using the following formula: $$S = -k \sum_i P_i \ln(P_i)$$

Here the summation is over the possible precisely defined states i the system can be in, and the P_i are the probabilities for the system to be in these states, given what we know about the system.

Robert Boyle

Robert Boyle (25 January 1627 - 30 December 1691) was an Irish theologian, natural philosopher, chemist, physicist, inventor, and early gentleman scientist, noted for his work in physics and chemistry. He is best known for the formulation of Boyle's law. Although his research and personal philosophy clearly has its roots in the alchemical tradition, he is largely regarded today as the first modern chemist, and therefore one of the founders of modern chemistry.

Caloric theory

The Caloric theory is an obsolete scientific theory that heat consists of a fluid called caloric that flows from hotter to colder bodies. Caloric was also thought of as a weightless gas that could pass in and out of pores in solids and liquids. The Caloric theory was superseded by the mid-19th century in favor of the theory of heat but nevertheless persisted in scientific literature until the end of the 19th century.

Watt

The Watt (symbol: W) is a derived unit of power in the International System of Units (SI.) It measures rate of energy conversion. One Watt is equivalent to 1 joule (J) of energy per second.

Wave

A Wave is a disturbance that propagates through space and time, usually with transference of energy. A mechanical Wave is a Wave that propagates or travels through a medium due to the restoring forces it produces upon deformation. There also exist Wave s capable of traveling through a vacuum, including electromagnetic radiation and probably gravitational radiation.

Waves

Waves is a three-part novel by Ogan Gurel published in 2009. A 21st century version of Faust, the novel explores good and evil in both individual and global settings focusing around a hypothetical technology that has both medical and military applications. The protagonist, Tomas Twarok, is a contemplative and idealistic doctor-turned-entrepreneur who strikes a deal with his college friend, Maximilian Iblis, a ruthless hedge fund manager.

Frequency

Frequency is the number of occurrences of a repeating event per unit time. It is also referred to as temporal Frequency The period is the duration of one cycle in a repeating event, so the period is the reciprocal of the Frequency

Internal energy

In thermodynamics, the Internal energy of a thermodynamic system denoted by U is the total of the kinetic energy due to the motion of molecules (translational, rotational, vibrational) and the potential energy associated with the vibrational and electric energy of atoms within molecules or crystals. It includes the energy in all of the chemical bonds, and the energy of the free, conduction electrons in metals.

One can also calculate the Internal energy of electromagnetic or blackbody radiation.

Kinetic energy

The Kinetic energy of an object is the extra energy which it possesses due to its motion. It is defined as the work needed to accelerate a body of a given mass from rest to its current velocity. Having gained this energy during its acceleration, the body maintains this Kinetic energy unless its speed changes.

Liquid

Liquid is one of the principal states of matter. A Liquid is a fluid that has the particles loose and can freely form a distinct surface at the boundaries of its bulk material. The surface is a free surface where the Liquid is not constrained by a container.

Matter

The term Matter traditionally refers to the substance that objects are made of. One common way to identify this 'substance' is through its physical properties; a common definition of Matter is anything that has mass and occupies a volume. However, this definition has to be revised in light of quantum mechanics, where the concept of 'having mass', and 'occupying space' are not as well-defined as in everyday life.

Oscillation

Oscillation is the repetitive variation, typically in time, of some measure about a central value or between two or more different states. Familiar examples include a swinging pendulum and AC power. The term vibration is sometimes used more narrowly to mean a mechanical oscillation but sometimes is used to be synonymous with 'oscillation.' Oscillations occur not only in physical systems but also in biological systems and in human society.

Potential energy

Potential energy can be thought of as energy stored within a physical system. It is called potential energy because it has the potential to be converted into other forms of energy, such as kinetic energy, and to do work in the process. The standard unit of measure for potential energy is the joule, the same as for work, or energy in general.

Radio	Radio is the transmission of signals, by modulation of electromagnetic waves with frequencies below those of visible light. Electromagnetic radiation travels by means of oscillating electromagnetic fields that pass through the air and the vacuum of space. Information is carried by systematically changing some property of the radiated waves, such as amplitude, frequency, or phase.
Radio waves	Radio waves are electromagnetic waves occurring on the radio frequency portion of the electromagnetic spectrum. A common use is to transport information through the atmosphere or outer space without wires. Radio waves are distinguished from other kinds of electromagnetic waves by their wavelength, a relatively long wavelength in the electromagnetic spectrum.
Seismic	Seismology is the scientific study of earthquakes and the propagation of elastic waves through the Earth. The field also includes studies of earthquake effects, such as tsunamis as well as diverse seismic sources such as volcanic, tectonic, oceanic, atmospheric, and artificial processes. A related field that uses geology to infer information regarding past earthquakes is paleoseismology.
Seismic wave	A seismic wave is a wave that travels through the Earth, most often as the result of a tectonic earthquake, sometimes from an explosion. Seismic waves are also continually excited by the pounding of ocean waves and the wind.
Seismic waves	Seismic waves are waves that travel through the Earth, most often as the result of a tectonic earthquake, sometimes from an explosion. Seismic waves are also continually excited by the pounding of ocean waves and the wind. Seismic waves are studied by seismologists, and measured by a seismograph, which records the output of a seismometer, or geophone.
Solid	The solid state of matter is characterized by a distinct structural rigidity and virtual resistance to deformation (i.e. changes of shape and/or volume.) Most solid s have high values both of Young's modulus and of the shear modulus of elasticity. This contrasts with most liquids or fluids, which have a low shear modulus, and typically exhibit the capacity for macroscopic viscous flow.
Vacuum	A vacuum is a volume of space that is essentially empty of matter, such that its gaseous pressure is much less than atmospheric pressure. The word comes from the Latin term for 'empty,' but in reality, no volume of space can ever be perfectly empty. A perfect vacuum with a gaseous pressure of absolute zero is a philosophical concept that is never observed in practice.
Vibration	Vibration refers to mechanical oscillations about an equilibrium point . The oscillations may be periodic such as the motion of a pendulum or random such as the movement of a tire on a gravel road. Vibration is occasionally 'desirable'.
X-radiation	X-radiation is a form of electromagnetic radiation. X-rays have a wavelength in the range of 10 to 0.01 nanometers, corresponding to frequencies in the range 30 petahertz to 30 exahertz and energies in the range 120 eV to 120 keV. They are longer than gamma rays but shorter than UV rays.
Ripple	The most common meaning of ripple in electrical science, is the small unwanted residual periodic variation of the direct current output of a power supply which has been derived from an alternating current source. This ripple is due to incomplete suppression of the alternating waveform within the power supply.

As well as this time-varying phenomenon, there is a frequency domain ripple that arises in some classes of filter and other signal processing networks.

Gas

In physics, a Gas is a state of matter, consisting of a collection of particles (molecules, atoms, ions, electrons, etc.) without a definite shape or volume that are in more or less random motion. Gas ses are generally modeled by the ideal Gas law where the total volume of the Gas increases proportionally to absolute temperature and decreases inversely proportionally to pressure.

Longitudinal wave

Longitudinal wave s are waves that have same direction of oscillations or vibrations along or parallel to their direction of travel, which means that the oscillations of the medium (particle) is in the same direction or opposite direction as the motion of the wave. Mechanical Longitudinal wave s have been also referred to as compressional waves or compression waves. Plane pressure wave Representation of the propagation of a Longitudinal wave on a 2d grid (empirical shape)

Examples of Longitudinal wave s include sound waves (alternation in pressure, particle displacement, or particle velocity propagated in an elastic material) and seismic P-waves (created by earthquakes and explosions.)

Transverse wave

A transverse wave is a moving wave that consists of oscillations occurring perpendicular to the direction of energy transfer. If a transverse wave is moving in the positive x-direction, its oscillations are in up and down directions that lie in the yz-plane. A transverse wave could be represented by moving a ribbon or piece of string, spread across a table, to the left and right or up and down.

Atmosphere

An Atmosphere is a layer of gases that may surround a material body of sufficient mass, by the gravity of the body, and are retained for a longer duration if gravity is high and the Atmosphere s temperature is low. Some planets consist mainly of various gases, but only their outer layer is their Atmosphere .

The term stellar Atmosphere describes the outer region of a star, and typically includes the portion starting from the opaque photosphere outwards.

Pulse

In physics, a *Pulse* is a single, abrupt emission of particles or radiation.

Density

The Density of a material is defined as its mass per unit volume. The symbol of Density is ρ '>rho.)

Mathematically:

$$\rho = \frac{m}{V}$$

where:

ρ (rho) is the Density
m is the mass,
V is the volume.

Linear

The word Linear comes from the Latin word Linear is, which means created by lines. In mathematics, a Linear map or function f(x) is a function which satisfies the following two properties...

· Additivity (also called the superposition property): $f(x + y) = f(x) + f(y.)$

Specular reflection

Specular reflection is the perfect, mirror-like reflection of light from a surface, in which light from a single incoming direction is reflected into a single outgoing direction. Such behavior is described by the law of reflection, which states that the direction of incoming light, and the direction of outgoing light reflected make the same angle with respect to the surface normal, thus the angle of incidence equals the angle of reflection; this is commonly stated as $>\theta_i = >\theta_r$. This behavior was first discovered through careful observation and measurement by Hero of Alexandria.>

This is in contrast to diffuse reflection, where incoming light is reflected in a broad range of directions.

Mass

Mass is a concept used in the physical sciences to explain a number of observable behaviours, and in everyday usage, it is common to identify Mass with those resulting behaviors. In particular, Mass is commonly identified with weight. But according to our modern scientific understanding, the weight of an object results from the interaction of its Mass with a gravitational field, so while Mass is part of the explanation of weight, it is not the complete explanation.

Air

The Earth's atmosphere is a layer of gases surrounding the planet Earth that is retained by Earth's gravity. The atmosphere protects life on Earth by absorbing ultraviolet solar radiation, warming the surface through heat retention (greenhouse effect), and reducing temperature extremes between day and night. Dry Air contains roughly (by volume) 78.08% nitrogen, 20.95% oxygen, 0.93% argon, 0.038% carbon dioxide, and trace amounts of other gases.

Temperature

Temperature is a physical property of a system that underlies the common notions of hot and cold; something that is hotter generally has the greater temperature. Specifically, temperature is a property of matter. Temperature is one of the principal parameters of thermodynamics.

Speed of sound

Sound is a vibration that travels through an elastic medium as a wave. The *Speed of sound* describes how much distance such a wave travels in a certain amount of time. In dry air at 20 °C (68 °F), the *Speed of sound* is 343 m/s.

Amplitude

Amplitude is the magnitude of change in the oscillating variable, with each oscillation, within an oscillating system. For instance, sound waves are oscillations in atmospheric pressure and their Amplitude s are proportional to the change in pressure during one oscillation. If the variable undergoes regular oscillations, and a graph of the system is drawn with the oscillating variable as the vertical axis and time as the horizontal axis, the Amplitude is visually represented by the vertical distance between the extrema of the curve.

Cycle

In combinatorial mathematics, a cycle of length n of a permutation P over a set S is a subset $\{ c_1, ..., c_n \}$ of S on which the permutation P acts in the following way:

$P(c_i) = c_{i+1}$ for $i = 1, ..., n - 1$, and $P(c_n) = c_1$.

It is usual to write a cycle by its mapping:

$(c_1\ c_2 \dots c_n .)$

Sine wave

The sine wave or sinusoid is a function that occurs often in mathematics, physics, signal processing, audition, electrical engineering, and many other fields. Its most basic form is:

$$y(t) = A \cdot \sin(\omega t + \theta)$$

which describes a wavelike function of time with:

· peak deviation from center = A
· angular frequency ω
· phase = θ

· When the phase is non-zero, the entire waveform appears to be shifted in time by the amount θ/ω seconds. A negative value represents a delay, and a positive value represents a 'head-start'.

The sine wave is important in physics because it retains its waveshape when added to another sine wave of the same frequency and arbitrary phase. It is the only periodic waveform that has this property. This property leads to its importance in Fourier analysis and makes it acoustically unique.

Wavelength

In physics wavelength is the distance between repeating units of a propagating wave of a given frequency. It is commonly designated by the Greek letter lambda. Examples of wave-like phenomena are light, water waves, and sound waves.

Shape

The shape of an object located in some space is the part of that space occupied by the object, as determined by its external boundary - abstracting from other properties such as colour, content, and material composition, as well as from the object's other spatial properties .

Mathematician and statistician David George Kendall defined shape this way:

Simple two-dimensional shape s can be described by basic geometry such as points, line, curves, plane, and so on.

Thermal

A thermal column (or thermal is a column of rising air in the lower altitudes of the Earth's atmosphere. thermal s are created by the uneven heating of the Earth's surface from solar radiation, and an example of convection. The Sun warms the ground, which in turn warms the air directly above it.

Waveform

Waveform means the shape and form of a signal such as a wave moving in a solid, liquid or gaseous medium.

In many cases the medium in which the wave is being propagated does not permit a direct visual image of the form. In these cases, the term 'waveform' refers to the shape of a graph of the varying quantity against time or distance.

Wave propagation

Wave propagation is any of the ways in which waves travel through a waveguide.

With respect to the direction of the oscillation relative to the propagation direction, we can distinguish between longitudinal wave and transverse waves.

For electromagnetic waves, propagation may occur in a vacuum as well as in a material medium.

Wavefront

In optics and physics, a wavefront is the locus of points having the same phase. Since infrared, optical, x-ray and gamma-ray frequencies are so high, the temporal component of electromagnetic waves is usually ignored at these wavelengths, and it is only the phase of the spatial oscillation that is described. Additionally, most optical systems and detectors are indifferent to polarization, so this property of the wave is also usually ignored.

Plane wave

In the physics of wave propagation, a Plane wave (also spelled planewave) is a constant-frequency wave whose wavefronts (surfaces of constant phase) are infinite parallel planes of constant amplitude normal to the phase velocity vector.

By extension, the term is also used to describe waves that are approximately Plane wave s in a localized region of space. For example, a localized source such as an antenna produces a field that is approximately a Plane wave in its far-field region.

Radiation

Radiation, as in physics, is energy in the form of waves or moving subatomic particles emitted by an atom or other body as it changes from a higher energy state to a lower energy state. Radiation can be classified as ionizing or non-ionizing radiation, depending on its effect on atomic matter. The most common use of the word 'radiation' refers to ionizing radiation.

Reflection

Reflection is the change in direction of a wave front at an interface between two different media so that the wave front returns into the medium from which it originated. Common examples include the reflection of light, sound and water waves.

Law of reflection: Angle of incidence = Angle of reflection

Reflections may occur in a number of wave and particle phenomena; these include acoustic, seismic waves in geologic structures, surface waves in bodies of water, and various electromagnetic waves, most usefully from VHF and higher radar frequencies, progressing upward through centimeter to millimeter-wavelength radar and the various light frequencies and (with special 'grazing' mirrors, to X-ray frequencies and beyond to gamma rays.

Satellite

In the context of spaceflight, a satellite is an object which has been placed into orbit by human endeavor. Such objects are sometimes called artificial satellite s to distinguish them from natural satellite s such as the Moon.

The first artificial satellite Sputnik 1, was launched by the Soviet Union in 1957.

Vesto Melvin Slipher

Vesto Melvin Slipher was an American astronomer. He used spectroscopy to investigate the rotation periods of planets, the composition of planetary atmospheres. In 1912, he was the first to observe the shift of spectral lines of galaxies.

Space

Space is the boundless, three-dimensional extent in which objects and events occur and have relative position and direction. Physical Space is often conceived in three linear dimensions, although modern physicists usually consider it, with time, to be part of the boundless four-dimensional continuum known as Space time. In mathematics Space s with different numbers of dimensions and with different underlying structures can be examined.

Telescope

A telescope is an instrument designed for the observation of remote objects by the collection of electromagnetic radiation. The first known practically functioning telescope s were invented in the Netherlands at the beginning of the 17th century. ' telescope s' can refer to a whole range of instruments operating in most regions of the electromagnetic spectrum.

Cosmology

Cosmology is the study of the Universe in its totality, and by extension, humanity's place in it. Though the word Cosmology is recent ' href='/wiki/Christian_Wolff_(philosopher)'>Christian Wolff's Cosmologia Generalis), study of the universe has a long history involving science, philosophy, esotericism, and religion.

In recent times, physics and astrophysics have played a central role in shaping the understanding of the universe through scientific observation and experiment; or what is known as physical Cosmology shaped through both mathematics and observation in the analysis of the whole universe.

Universe

The Universe is defined as everything that physically exists: the entirety of space and time, all forms of matter, energy and momentum, and the physical laws and constants that govern them. However, the term 'universe' may be used in slightly different contextual senses, denoting such concepts as the cosmos, the world or Nature.

Astronomical observations indicate that the universe is 13.73 ± 0.12 billion years old and at least 93 billion light years across.

Astronomy

Astronomy , 'law') is the scientific study of celestial objects (such as stars, planets, comets, and galaxies) and phenomena that originate outside the Earth's atmosphere (such as the cosmic background radiation.) It is concerned with the evolution, physics, chemistry, meteorology, and motion of celestial objects, as well as the formation and development of the universe.

Radar

Radar is a system that uses electromagnetic waves to identify the range, altitude, direction, or speed of both moving and fixed objects such as aircraft, ships, motor vehicles, weather formations, and terrain. The term RADAR was coined in 1941 as an acronym for Radio Detection and Ranging. The term has since entered the English language as a standard word, radar, losing the capitalization.

Sonar

Sonar is a technique that uses sound propagation to navigate, communicate or to detect other vessels. There are two kinds of sonar--active and passive. Sonar may be used as a means of acoustic location.

Star

A star is a massive, luminous ball of plasma that is held together by gravity. The nearest star to Earth is the Sun, which is the source of most of the energy on Earth. Other star s are visible in the night sky, when they are not outshone by the Sun.

Diffraction

Diffraction is normally taken to refer to various phenomena which occur when a wave encounters an obstacle. It is described as the apparent bending of waves around small obstacles and the spreading out of waves past small openings. Similar effects are observed when light waves travel through a medium with a varying refractive index or a sound wave through one with varying acoustic impedance.

Shock

A mechanical or physical Shock is a sudden acceleration or deceleration caused, for example, by impact, drop, kick, earthquake, or explosion. Shock is a transient physical excitation.

Shock is usually measured by an accelerometer.

Shock wave

A shock wave is a type of propagating disturbance. Like an ordinary wave, it carries energy and can propagate through a medium or in some cases in the absence of a material medium, through a field such as the electromagnetic field. shock wave s are characterized by an abrupt, nearly discontinuous change in the characteristics of the medium.

Sonic boom	The term Sonic boom is commonly used to refer to the shocks caused by the supersonic flight of an aircraft. Sonic boom s generate enormous amounts of sound energy, sounding much like an explosion. Thunder is a type of natural Sonic boom created by the rapid heating and expansion of air in a lightning discharge.
Atmospheric pressure	Atmospheric pressure is sometimes defined as the force per unit area exerted against a surface by the weight of air above that surface at any given point in the Earth's atmosphere. In most circumstances Atmospheric pressure is closely approximated by the hydrostatic pressure caused by the weight of air above the measurement point. Low pressure areas have less atmospheric mass above their location, whereas high pressure areas have more atmospheric mass above their location.
Molecule	A Molecule is defined as a sufficiently stable, electrically neutral group of at least two atoms in a definite arrangement held together by very strong (covalent) chemical bonds. Molecule s are distinguished from polyatomic ions in this strict sense. In organic chemistry and biochemistry, the term Molecule is used less strictly and also is applied to charged organic Molecule s and bio Molecule s.
Pressure waves	Pressure waves are one of the types of elastic waves, also called seismic waves, that can travel through elastic solids, including the Earth. The waves can be produced by earthquakes and recorded by seismometers.
Infrasound	Infrasound is sound that is lower in frequency than 20 cycles per second, the normal limit of human hearing. Hearing becomes gradually less sensitive as frequency decreases, so for humans to perceive Infrasound the sound pressure must be sufficiently high. The ear is the primary organ for sensing Infrasound but at higher levels it is possible to feel Infrasound vibrations in various parts of the body.
Noise	Noise in audio, recording, and broadcast systems refers to the residual low level sound (usually hiss and hum) that is heard in quiet periods of a programme. In audio engineering it can refer either to the acoustic Noise from loudspeakers, or to the unwanted residual electronic Noise signal that gives rise to acoustic Noise heard as 'hiss'. This signal Noise is commonly measured using A-weighting or ITU-R 468 weighting Noise is often generated deliberately and used as a test signal.
Oscilloscope	An oscilloscope is a type of electronic test instrument that allows signal voltages to be viewed, usually as a two-dimensional graph of one or more electrical potential differences (vertical axis) plotted as a function of time or of some other voltage (horizontal axis.) Although an oscilloscope displays voltage on its vertical axis, any other quantity that can be converted to a voltage can be displayed as well. In most instances, oscilloscope s show events that repeat with either no change, or change slowly.
Pure tone	A pure tone is a tone with a sinusoidal waveshape. A sine wave is characterized by its frequency -- the number of cycles per second, or its wavelength -- the distance the waveform travels through its medium within a period, and the amplitude -- the size of each cycle. A pure tone has the unique property that its waveshape and sound are changed only in amplitude and phase by linear acoustic systems.

Ultrasound

Ultrasound is cyclic sound pressure with a frequency greater than the upper limit of human hearing. Although this limit varies from person to person, it is approximately 20 kilohertz in healthy, young adults and thus, 20 kHz serves as a useful lower limit in describing ultrasound. The production of ultrasound is used in many different fields, typically to penetrate a medium and measure the reflection signature or supply focused energy.

Ton

There are several similar units of mass or volume called the ton

Others

· The long ton is used for petroleum products such as aviation fuel.
· Deadweight ton (abbreviation 'DWT' or 'dwt') is a measure of a ship's carrying capacity, including bunker oil, fresh water, ballast water, crew and provisions. It is expressed in ton nes (1000 kg) or long ton s (2240 pounds, about 1016 kg.) This measurement is also used in the U.S. ton nage of naval ships.
· Increasingly, ton nes are being used rather than long ton s in measuring the displacement of ships. See ton nage.
· Harbour ton used in South Africa in the 20th century, 2000 pounds or one short ton

Both the long ton and the short ton are composed of 20 hundredweight, being 112 and 100 pounds respectively. Prior to the 15th century in England, the ton was composed of 20 hundredweight, each of 108 lb, giving a ton of 2160 pounds.
Assay ton (abbreviation 'AT') is not a unit of measurement, but a standard quantity used in assaying ores of precious metals; it is $29\frac{1}{6}$ grams (short assay ton or $32\frac{2}{3}$ grams (long assay ton , the amount which bears the same ratio to a milligram as a short or long ton bears to a troy ounce. In other words, the number of milligrams of a particular metal found in a sample of this size gives the number of troy ounces contained in a short or long ton of ore.

Meteorology

Meteorology is the interdisciplinary scientific study of the atmosphere that focuses on weather processes and forecasting . Studies in the field stretch back millennia, though significant progress in Meteorology did not occur until the eighteenth century. The nineteenth century saw breakthroughs occur after observing networks developed across several countries.

Radiosonde

A Radiosonde is a unit for use in weather balloons that measures various atmospheric parameters and transmits them to a fixed receiver. Radiosonde s may operate at a radio frequency of 403 MHz or 1680 MHz and both types may be adjusted slightly higher or lower as required. A rawinsonde is a Radiosonde that is designed to also measure wind speed and direction.

Velocity

In physics, velocity is defined as the rate of change of position. It is a vector physical quantity; both speed and direction are required to define it. In the SI system, it is measured in meters per second: or ms^{-1}.

Wind shear

Wind shear, sometimes referred to as windshear or wind gradient, is a difference in wind speed and direction over a relatively short distance in the atmosphere. Wind shear can be broken down into vertical and horizontal components, with horizontal wind shear seen across weather fronts and near the coast, and vertical shear typically near the surface, though also at higher levels in the atmosphere near upper level jets and frontal zones aloft.

Wind shear itself is a microscale meteorological phenomenon occurring over a very small distance, but it can be associated with mesoscale or synoptic scale weather features such as squall lines and cold fronts.

Wind speed

Wind speed is the speed of wind, the movement of air or other gases in an atmosphere. It is a scalar quantity, the magnitude of the vector of motion.

Wind speed has always meant the movement of air in an outside environment, but the speed of air movement inside is important in many areas, including weather forecasting, aircraft and maritime operations, building and civil engineering.

Radio acoustic sounding system

A radio acoustic sounding system is a system for measuring the atmospheric lapse rate using backscattering of radio waves from an acoustic wave front to measure the speed of sound at various heights above the ground. This is possible because the compression and rarefaction of air by an acoustic wave changes the dielectric properties, producing partial reflection of the transmitted radar signal. From the speed of sound, the temperature of the air in the planetary boundary layer can be computed.

Refrigerator

A Refrigerator (often called a 'fridge' for short) is a cooling appliance comprising a thermally insulated compartment and a heat pump--chemical or mechanical means--to transfer heat from it to the external environment, cooling the contents to a temperature below ambient. Refrigerator s are extensively used to store foods which spoil from bacterial growth if not refrigerated. A device described as a Refrigerator maintains a temperature a few degrees above the freezing point of water; a similar device which maintains a temperature below the freezing point of water is called a 'freezer.' The Refrigerator is a relatively modern invention among kitchen appliances.

Sodar

SODAR is a meteorological instrument also known as a wind profiler which measures the scattering of sound waves by atmospheric turbulence. SODAR systems are used to measure wind speed at various heights above the ground, and the thermodynamic structure of the lower layer of the atmosphere.

Sodar systems are like radar systems except that sound waves rather than radio waves are used for detection.

Acoustic

Acoustic or sonic lubrication occurs when sound (measurable in a vacuum by placing a microphone on one element of the sliding system) permits vibration to introduce separation between the sliding faces. This could happen between two plates or between a series of particles. The frequency of sound required to induce optimal vibration, and thus cause sonic lubrication, varies with the size of the particles (high frequencies will have the desired, or undesired, effect on sand and lower frequencies will have this effect on boulders.)

Ranging

Ranging is a process or method to determine the distance from one location or position to another location or position. Another term for this method is lateration, see unilateration. Further development led to cooperative systems, where both locations are equipped with respective apparatus and thus provide bilateral measuring and to multilateral measurement between a larger set of locations, see multilateration.

Sonoluminescence

Sonoluminescence is the emission of short bursts of light from imploding bubbles in a liquid when excited by sound.

The effect was first discovered at the University of Cologne in 1934 as a result of work on sonar. H. Frenzel and H. Schultes put an ultrasound transducer in a tank of photographic developer fluid.

Reverberation

Reverberation is the persistence of sound in a particular space after the original sound is removed. When sound is produced in a space, a large number of echoes build up and then slowly decay as the sound is absorbed by the walls and air, creating reverberation, or reverb. This is most noticeable when the sound source stops but the reflections continue, decreasing in amplitude, until they can no longer be heard.

Loudness

Loudness is the quality of a sound that is the primary psychological correlate of physical strength (amplitude.) The horizontal axis shows frequency in Hz

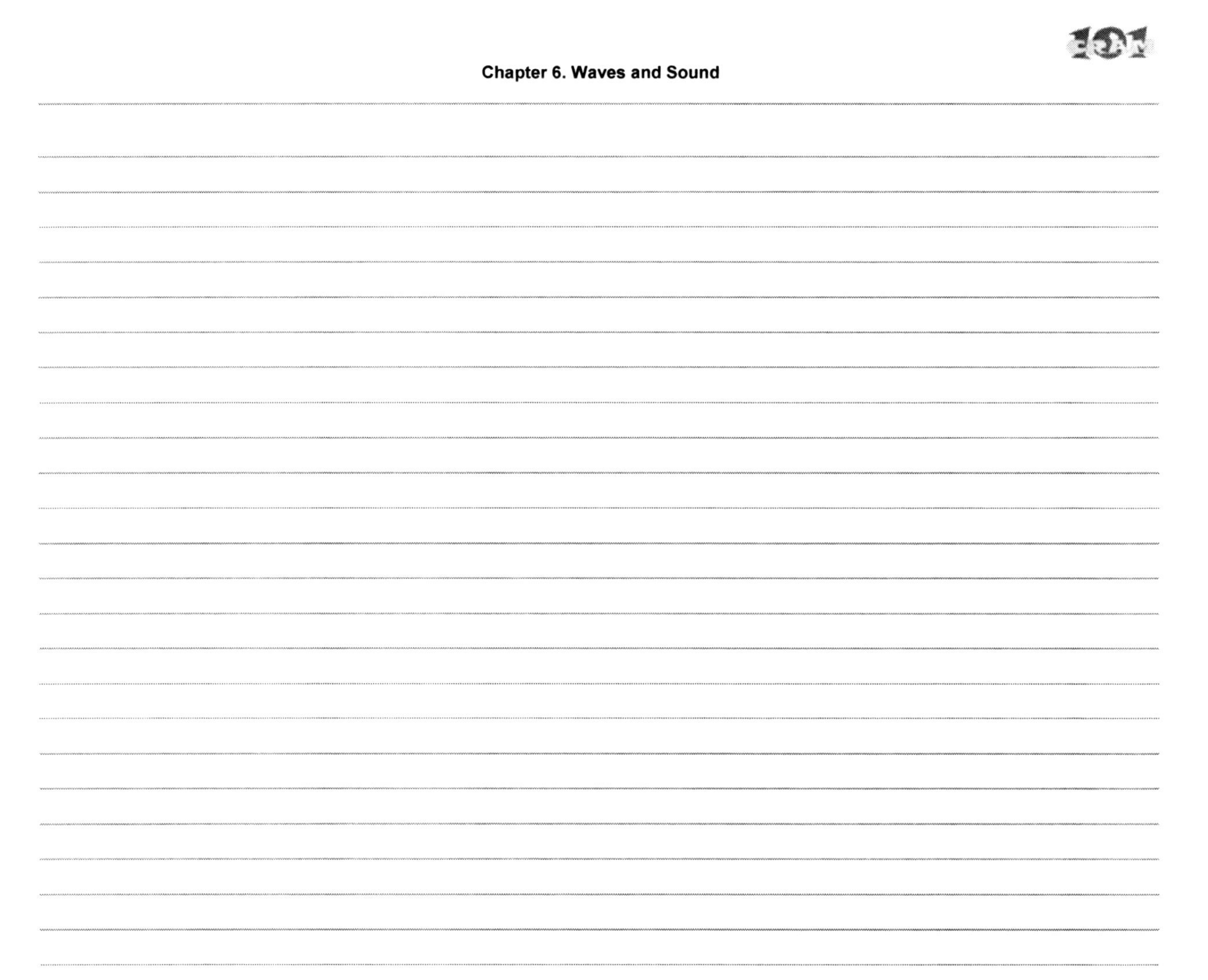

Pitch
Pitch is the name for any of a number of highly viscous liquids which appear solid. Pitch can be made from petroleum products or plants. Petroleum-derived Pitch is also called bitumen.

Tone quality
In music, timbre is the quality of a musical note or sound that distinguishes different types of sound production, such as voices or musical instruments. The physical characteristics of sound that mediate the perception of timbre include spectrum and envelope. Timbre is also known in psychoacoustics as tone quality or tone color.

Decibel
The Decibel (dB) is a logarithmic unit of measurement that expresses the magnitude of a physical quantity (usually power or intensity) relative to a specified or implied reference level. Since it expresses a ratio of two quantities with the same unit, it is a dimensionless unit. A Decibel is one tenth of a bel, a seldom-used unit.

Harmonic
In acoustics and telecommunication, a Harmonic of a wave is a component frequency of the signal that is an integer multiple of the fundamental frequency. For example, if the fundamental frequency is f, the Harmonic s have frequencies f, 2f, 3f, 4f, etc. The Harmonic s have the property that they are all periodic at the fundamental frequency, therefore the sum of Harmonic s is also periodic at that frequency.

White
White is not a color, the perception which is evoked by light that stimulates all three types of color sensitive cone cells in the human eye in near equal amount and with high brightness compared to the surroundings.

Since the impression of white is obtained by three summations of light intensity across the visible spectrum, the number of combinations of light wavelengths that produce the sensation of white is practically infinite. There are a number of different white light sources such as the midday Sun, incandescent lamps, fluorescent lamps and white LEDs.

White light
White is a tone, the perception which is evoked by light that stimulates all three types of color sensitive cone cells in the human eye in near equal amount and with high brightness compared to the surroundings.

Since the impression of white is obtained by three summations of light intensity across the visible spectrum, the number of combinations of light wavelengths that produce the sensation of white is practically infinite. There are a number of different white light sources such as the midday Sun, incandescent lamps, fluorescent lamps and white LEDs.

White noise
White noise is a random signal with a flat power spectral density. In other words, the signal contains equal power within a fixed bandwidth at any center frequency. White noise draws its name from white light in which the power spectral density of the light is distributed over the visible band in such a way that the eye's three color receptors are rather equally stimulated.

Acoustics
Acoustics is the interdisciplinary science that deals with the study of sound, ultrasound and infrasound (all mechanical waves in gases, liquids, and solids.) A scientist who works in the field of Acoustics is an acoustician. The application of Acoustics in technology is called acoustical engineering.

Ohm
The Ohm (symbol: Ω) is the SI unit of electrical impedance or, in the direct current case, electrical resistance applied to these points, produces in the conductor a current of 1 ampere, the conductor not being the seat of any electromotive force.

$$\Omega = \frac{V}{A} = \frac{m^2 \cdot kg}{s \cdot C^2} = \frac{J}{s \cdot A^2} = \frac{kg \cdot m^2}{s^3 \cdot A^2} = \frac{HP}{745.69 \cdot A^2}$$

In many cases the resistance of a conductor in Ohm s is approximately constant within a certain range of voltages, temperatures, and other parameters; one speaks of linear resistors. In other cases resistance varies (e.g., thermistors.)

Optics

Optics is the science that describes the behavior and properties of light and the interaction of light with matter. Optics explains optical phenomena. The word optics comes from á½€πτικÎ®, meaning appearance or look in Ancient Greek.)

Pythagoras

Pythagoras of Samos was an Ionian Greek mathematician and founder of the religious movement called Pythagoreanism. He is often revered as a great mathematician, mystic and scientist; however some have questioned the scope of his contributions to mathematics and natural philosophy. Herodotus referred to him as 'the most able philosopher among the Greeks'.

Conservation law

In physics, a Conservation law states that a particular measurable property of an isolated physical system does not change as the system evolves.

One particularly important physical result concerning Conservation law s is Noether's Theorem, which states that there is a one-to-one correspondence between Conservation law s and differentiable symmetries of physical systems. For example, the conservation of energy follows from the time-invariance of physical systems, and the fact that physical systems behave the same regardless of how they are oriented in space gives rise to the conservation of angular momentum.

Electricity

Electricity is a general term that encompasses a variety of phenomena resulting from the presence and flow of electric charge. These include many easily recognizable phenomena, such as lightning and static Electricity but in addition, less familiar concepts, such as the electromagnetic field and electromagnetic induction.

In general usage, the word Electricity is adequate to refer to a number of physical effects.

Magnetism

In physics, the term Magnetism is used to describe how materials respond on the microscopic level to an applied magnetic field; to categorize the magnetic phase of a material. For example, the most well known form of Magnetism is ferro Magnetism such that some ferromagnetic materials produce their own persistent magnetic field. Some well-known ferromagnetic materials that exhibit easily detectable magnetic properties (to form magnets) are nickel, iron, cobalt, gadolinium and their alloys.

Quantum

In physics, a quantum is an indivisible entity of a quantity that has the same units as the Planck constant and is related to both energy and momentum of elementary particles of matter and of photons and other bosons. The word comes from the Latin 'quantus,' for 'how much.' Behind this, one finds the fundamental notion that a physical property may be 'quantized', referred to as 'quantization'. This means that the magnitude can take on only certain discrete numerical values, rather than any value, at least within a range.

Quantum mechanics

Quantum mechanics is the study of mechanical systems whose dimensions are close to the atomic scale, such as molecules, atoms, electrons, protons and other subatomic particles. Quantum mechanics is a fundamental branch of physics with wide applications. Quantum theory generalizes classical mechanics to provide accurate descriptions for many previously unexplained phenomena such as black body radiation and stable electron orbits.

Electric charge	Electric charge is a fundamental conserved property of some subatomic particles, which determines their electromagnetic interaction. Electrically charged matter is influenced by, and produces, electromagnetic fields. The interaction between a moving charge and an electromagnetic field is the source of the electromagnetic force, which is one of the four fundamental forces.
Electric field	In physics, the space surrounding an electric charge or in the presence of a time-varying magnetic field has a property called an Electric field This Electric field exerts a force on other electrically charged objects. The concept of an Electric field was introduced by Michael Faraday.
Electricity	Electricity is a general term that encompasses a variety of phenomena resulting from the presence and flow of electric charge. These include many easily recognizable phenomena, such as lightning and static Electricity but in addition, less familiar concepts, such as the electromagnetic field and electromagnetic induction. In general usage, the word Electricity is adequate to refer to a number of physical effects.
Electrostatic	Electrostatic s is the branch of science that deals with the phenomena arising from stationary or slowly moving electric charges. Since classical antiquity it was known that some materials such as amber attract light particles after rubbing. The Greek word for amber, Î®λεκτρον , was the source of the word 'electricity'.
Ionosphere	The Ionosphere is the uppermost part of the atmosphere, distinguished because it is ionized by solar radiation. It plays an important part in atmospheric electricity and forms the inner edge of the magnetosphere. It has practical importance because, among other functions, it influences radio propagation to distant places on the Earth.
Liquid	Liquid is one of the principal states of matter. A Liquid is a fluid that has the particles loose and can freely form a distinct surface at the boundaries of its bulk material. The surface is a free surface where the Liquid is not constrained by a container.
Liquid crystals	Liquid crystals (Liquid crystals s) are substances that exhibit a phase of matter that has properties between those of a conventional liquid and those of a solid crystal. For instance, an Liquid crystals may flow like a liquid, but its molecules may be oriented in a crystal-like way. There are many different types of Liquid crystals phases, which can be distinguished by their different optical properties (such as birefringence.)
Transistor	In electronics, a transistor is a semiconductor device commonly used to amplify or switch electronic signals. A transistor is made of a solid piece of a semiconductor material, with at least three terminals for connection to an external circuit. A voltage or current applied to one pair of the transistor's terminals changes the current flowing through another pair of terminals.
Atom	The Atom is a basic unit of matter consisting of a dense, central nucleus surrounded by a cloud of negatively charged electrons. The Atom ic nucleus contains a mix of positively charged protons and electrically neutral neutrons (except in the case of hydrogen-1, which is the only stable nuclide with no neutron.) The electrons of an Atom are bound to the nucleus by the electromagnetic force.

Big Bang

The Big Bang is the cosmological model of the initial conditions and subsequent development of the universe that is supported by the most comprehensive and accurate explanations from current scientific evidence and observation. As used by cosmologists, the term Big Bang generally refers to the idea that the universe has expanded from a primordial hot and dense initial condition at some finite time in the past (currently estimated to have been approximately 13.7 billion years ago), and continues to expand to this day.

Georges Lemaître proposed what became known as the Big Bang theory of the origin of the Universe, although he called it his 'hypothesis of the primeval atom'.

Elementary particle

In particle physics, an Elementary particle or fundamental particle is a particle not known to have substructure; that is, it is not known to be made up of smaller particles. If an Elementary particle truly has no substructure, then it is one of the basic building blocks of the universe from which all other particles are made. In the Standard Model, the quarks, leptons, and gauge bosons are Elementary particle s.

Thermal neutron

A thermal neutron is a free neutron that is Boltzmann distributed with kT = 0.024 eV (4.0×10^{-21} J) at room temperature. This gives characteristic (not average, or median) speed of 2.2 km/s. The name 'thermal' comes from their energy being that of the room temperature gas or material they are permeating.

Proton

The Proton is a subatomic particle with an electric charge of +1 elementary charge. It is found in the nucleus of each atom but is also stable by itself and has a second identity as the hydrogen ion, $^{1}H^{+}$. It is composed of three even more fundamental particles comprising two up quarks and one down quark.

Subatomic

A subatomic particle is an elementary or composite particle smaller than an atom. Particle physics and nuclear physics are concerned with the study of these particles, their interactions, and non-atomic matter.

subatomic particles include the atomic constituents electrons, protons, and neutrons.

Subatomic particle

A subatomic particle is an elementary or composite particle smaller than an atom. Particle physics and nuclear physics are concerned with the study of these particles, their interactions, and non-atomic matter.

Subatomic particles include the atomic constituents electrons, protons, and neutrons.

Subatomic particles

In physics, subatomic particles are the particles composing nucleons and atoms. There are two types of subatomic particles elementary particles, which are not made of other particles, and composite particles. Particle physics and nuclear physics study these particles and how they interact.

Positive

In astrology, a positive dominant, active, yang, diurnal or masculine sign refers to any of the six odd-numbered signs of the zodiac: Aries, Gemini, Leo, Libra, Sagittarius or Aquarius.

These signs constitute the fire and air triplicities. They are termed positive because they tend to be more communicative, buoyant and sociable - broadly matching the extrovert personality, aiming for a large number of friends and to focusing on working with them rather than on their own resources.

Static

Statics is the branch of mechanics concerned with the analysis of loads on physical systems in static equilibrium, that is, in a state where the relative positions of subsystems do not vary over time, or where components and structures are at rest. When in static equilibrium, the system is either at rest, or its center of mass moves at constant velocity.

By Newton's second law, this situation implies that the net force and net torque on every body in the system is zero, meaning that for every force bearing upon a member, there must be an equal and opposite force.

Static cling

Static Cling is caused by static electricity, usually due to rubbing as in a clothes dryer. It can be removed by deionizing materials with water, and prevented with fabric softener dryer sheets. Antistatic agents are used to make the surfaces slightly conductive, which reduces or prevents the static charge buildup.

Force

In physics, a Force is any external agent that causes a change in the motion of a free body, or that causes stress in a fixed body. It can also be described by intuitive concepts such as a push or pull that can cause an object with mass to change its velocity , i.e., to accelerate, or which can cause a flexible object to deform. Force has both magnitude and direction, making it a vector quantity.

Matter

The term Matter traditionally refers to the substance that objects are made of. One common way to identify this 'substance' is through its physical properties; a common definition of Matter is anything that has mass and occupies a volume. However, this definition has to be revised in light of quantum mechanics, where the concept of 'having mass', and 'occupying space' are not as well-defined as in everyday life.

Orbit

In physics, an orbit is the gravitationally curved path of one object around a point or another body, for example the gravitational orbit of a planet around a star.

Historically, the apparent motion of the planets were first understood in terms of epicycles, which are the sums of numerous circular motions. This predicted the path of the planets quite well, until Johannes Kepler was able to show that the motion of the planets were in fact elliptical motions.

Orbits

The velocity relationship of two objects with mass can thus be considered in four practical classes, with subtypes:

· No orbit
· Interrupted orbits

· Range of interrupted elliptical paths
· Circumnavigating orbits

· Range of elliptical paths with closest point opposite firing point
· Circular path
· Range of elliptical paths with closest point at firing point
· Infinite orbits

· Parabolic paths
· Hyperbolic paths

In many situations relativistic effects can be neglected, and Newton's laws give a highly accurate description of the motion. Then the acceleration of each body is equal to the sum of the gravitational forces on it, divided by its mass, and the gravitational force between each pair of bodies is proportional to the product of their masses and decreases inversely with the square of the distance between them. To this Newtonian approximation, for a system of two point masses or spherical bodies, only influenced by their mutual gravitation, the orbits can be exactly calculated. If the heavier body is much more massive than the smaller, as for a satellite or small moon orbiting a planet or for the Earth orbiting the Sun, it is accurate and convenient to describe the motion in a coordinate system that is centered on the heavier body, and we can say that the lighter body is in orbit around the heavier.

Salt

Salt is a dietary mineral composed primarily of sodium chloride that is essential for animal life, but toxic to most land plants. Salt flavor is one of the basic tastes, and salt is the most popular food seasoning. Salt is also an important preservative.

Salt for human consumption is produced in different forms: unrefined salt, refined salt, and iodized salt.

Sodium

Sodium is an element which has the symbol N, atomic number 11, atomic mass 22.9898 g/mol, common oxidation number +1. Sodium is a soft, silvery white, highly reactive element and is a member of the alkali metals within 'group 1'. It has only one stable isotope, ^{23}Na.

Universal gravitation

Newton's law of universal gravitation is a physical law describing the gravitational attraction between bodies with mass. It is a part of classical mechanics and was first formulated in Newton's work Philosophiae Naturalis Principia Mathematica, first published on July 5, 1687. In modern language it states the following:

Every point mass attracts every other point mass by a force pointing along the line intersecting both points.

Conservation law

In physics, a Conservation law states that a particular measurable property of an isolated physical system does not change as the system evolves.

One particularly important physical result concerning Conservation law s is Noether's Theorem, which states that there is a one-to-one correspondence between Conservation law s and differentiable symmetries of physical systems. For example, the conservation of energy follows from the time-invariance of physical systems, and the fact that physical systems behave the same regardless of how they are oriented in space gives rise to the conservation of angular momentum.

Electron

An Electron is a subatomic particle that carries a negative electric charge. It has no known substructure and is believed to be a point particle. An Electron has a mass that is approximately 1836 times less than that of the proton.

Molecule

A Molecule is defined as a sufficiently stable, electrically neutral group of at least two atoms in a definite arrangement held together by very strong (covalent) chemical bonds. Molecule s are distinguished from polyatomic ions in this strict sense. In organic chemistry and biochemistry, the term Molecule is used less strictly and also is applied to charged organic Molecule s and bio Molecule s.

Polarity

In chemistry, polarity refers to the dipole-dipole intermolecular forces between the slightly positively-charged end of one molecule to the negative end of another or the same molecule. Molecular polarity is dependent on the difference in electronegativity between atoms in a compound and the asymmetry of the compound's structure. For example, water is thought to be polar because of the unequal sharing of its electrons.

Collision

A Collision is an isolated event in which two or more moving bodies (colliding bodies) exert relatively strong forces on each other for a relatively short time. Deflection happens when an object hits a plane surface

Collision s involve forces (there is a change in velocity.) Collision s can be elastic, meaning they conserve energy and momentum, inelastic, meaning they conserve momentum but not energy, or totally inelastic (or plastic), meaning they conserve momentum and the two objects stick together.

Humidity

Humidity is the amount of water vapour in the air. In daily language the term Humidity is normally taken to mean relative Humidity Relative Humidity is defined as the ratio of the partial pressure of water vapour in a parcel of air to the saturated vapour pressure of water vapour at a prescribed temperature.

Lightning

Lightning is an atmospheric discharge of electricity accompanied by thunder, which typically occurs during thunderstorms, and sometimes during volcanic eruptions or dust storms. In the atmospheric electrical discharge, a leader of a bolt of Lightning can travel at speeds of 60,000 m/s (130,000 mph), and can reach temperatures approaching 30,000 °C (54,000 °F), hot enough to fuse silica sand into glass channels known as fulgurites which are normally hollow and can extend some distance into the ground. There are some 16 million Lightning storms in the world every year.

Relative humidity

Relative humidity is a measurement of the amount of water vapor that exists in a gaseous mixture of air and water.

The relative humidity of an air-water vapor mixture can be estimated if both the temperature and the dew point temperature of the mixture are known. When both T and T_d are expressed in degrees celsius then :

$$RH = \frac{e_p}{e_s} \times 100\%$$

where the partial pressure of water vapor in the mixture is estimated by e_p :

$$e_p = e^{\frac{(17.269 \times T_d)}{(273.3 + T_d)}}$$

and the saturated vapor pressure of water at the temperature of the mixture is estimated by e_s :

$$e_s = e^{\frac{(17.269 \times T)}{(273.3 + T)}}$$

In practice both T and T_d are readily estimated by using a sling psychrometer and the relative humidity of the atmosphere can be calculated

Often the notion of air holding water vapor is used to describe the concept of relative humidity.

Spark

In mathematics, specifically in linear algebra, the Spark of a matrix A is the smallest number n such that there exists a set of n columns in A which are linearly dependent. Formally,

$$\text{spark}(A) = \min_{d \neq 0} \|d\|_0 \text{ s.t. } Ad = 0.$$

By contrast, the rank of a matrix is the smallest number k such that all sets of k + 1 columns in A are linearly dependent. The concept of the Spark is of use in the theory of compressive sensing, where requirements on the Spark of the measurement matrix are used to ensure stability and consistency of various estimation techniques.

Air

The Earth's atmosphere is a layer of gases surrounding the planet Earth that is retained by Earth's gravity. The atmosphere protects life on Earth by absorbing ultraviolet solar radiation, warming the surface through heat retention (greenhouse effect), and reducing temperature extremes between day and night. Dry Air contains roughly (by volume) 78.08% nitrogen, 20.95% oxygen, 0.93% argon, 0.038% carbon dioxide, and trace amounts of other gases.

Pollution

Pollution is the introduction of contaminants into an environment that causes instability, disorder, harm or discomfort to the physical systems or living organisms they are in. Pollution can take the form of chemical substances, or energy, such as noise, heat, or light energy. Pollutants, the elements of pollution, can be foreign substances or energies, or naturally occurring; when naturally occurring, they are considered contaminants when they exceed natural levels.

Ohm

The Ohm (symbol: Ω) is the SI unit of electrical impedance or, in the direct current case, electrical resistance applied to these points, produces in the conductor a current of 1 ampere, the conductor not being the seat of any electromotive force.

$$\Omega = \frac{\mathrm{V}}{\mathrm{A}} = \frac{\mathrm{m}^2 \cdot \mathrm{kg}}{\mathrm{s} \cdot \mathrm{C}^2} = \frac{\mathrm{J}}{\mathrm{s} \cdot \mathrm{A}^2} = \frac{\mathrm{kg} \cdot \mathrm{m}^2}{\mathrm{s}^3 \cdot \mathrm{A}^2} = \frac{\mathrm{HP}}{745.69 \cdot \mathrm{A}^2}$$

In many cases the resistance of a conductor in Ohm s is approximately constant within a certain range of voltages, temperatures, and other parameters; one speaks of linear resistors. In other cases resistance varies (e.g., thermistors.)

Semiconductor

A semiconductor is a material that has an electrical resistivity between that of a conductor and an insulator. An external electrical field changes a semiconductor s resistivity. Devices made from semiconductor materials are the foundation of modern electronics, including radio, computers, telephones, and many other devices.

Electromagnetism

Electromagnetism is the physics of the electromagnetic field, a field that exerts a force on particles with the property of electric charge and is reciprocally affected by the presence and motion of such particles.

A changing magnetic field produces an electric field (this is the phenomenon of electromagnetic induction, the basis of operation for electrical generators, induction motors, and transformers.) Similarly, a changing electric field generates a magnetic field.

Heat

In physics and thermodynamics, Heat is the process of energy transfer from one body or system due to thermal contact, which in turn is defined as an energy transfer to a body in any other way than due to work performed on the body.

When an infinitesimal amount of Heat δQ is tranferred to a body in thermal equilibrium at absolute temperature T in a reversible way, then it is given by the quantity TdS, where S is the entropy of the body.

A related term is thermal energy, loosely defined as the energy of a body that increases with its temperature.

Magnetism	In physics, the term Magnetism is used to describe how materials respond on the microscopic level to an applied magnetic field; to categorize the magnetic phase of a material. For example, the most well known form of Magnetism is ferro Magnetism such that some ferromagnetic materials produce their own persistent magnetic field. Some well-known ferromagnetic materials that exhibit easily detectable magnetic properties (to form magnets) are nickel, iron, cobalt, gadolinium and their alloys.
Nitrogen	Nitrogen is a chemical element that has the symbol N and atomic number 7 and atomic weight 14.0067. Elemental nitrogen is a colorless, odorless, tasteless and mostly inert diatomic gas at standard conditions, constituting 80% by volume of Earth's atmosphere. Many industrially important compounds, such as ammonia, nitric acid, organic nitrates, and cyanides, contain nitrogen.
Heike Kamerlingh Onnes	Heike Kamerlingh Onnes (21 September 1853 - 21 February 1926) was a Dutch physicist and Nobel laureate. His scientific career was spent exploring extremely cold refrigeration techniques and the associated phenomena. Kamerlingh Onnes was born in Groningen, Netherlands.
Particle accelerator	A particle accelerator is a device that uses electric fields to propel electrically-charged particles to high speeds and to contain them. An ordinary CRT television set is a simple form of accelerator. There are two basic types: linear accelerators and circular accelerators.
Temperature	Temperature is a physical property of a system that underlies the common notions of hot and cold; something that is hotter generally has the greater temperature. Specifically, temperature is a property of matter. Temperature is one of the principal parameters of thermodynamics.
Electromagnet	An Electromagnet is a type of magnet in which the magnetic field is produced by the flow of electric current. The magnetic field disappears when the current ceases. Electromagnet attracts paper clips when current is applied creating a magnetic field, loses them when current and magnetic field are removed A wire with an electric current passing through it, generates a magnetic field around it, this is a simple Electromagnet
Resonance	In physics, resonance is the tendency of a system to oscillate at maximum amplitude at certain frequencies, known as the system's resonance frequencies. At these frequencies, even small periodic driving forces can produce large amplitude vibrations, because the system stores vibrational energy. When damping is small, the resonance frequency is approximately equal to the natural frequency of the system, which is the frequency of free vibrations.
Internal energy	In thermodynamics, the Internal energy of a thermodynamic system denoted by U is the total of the kinetic energy due to the motion of molecules (translational, rotational, vibrational) and the potential energy associated with the vibrational and electric energy of atoms within molecules or crystals. It includes the energy in all of the chemical bonds, and the energy of the free, conduction electrons in metals. One can also calculate the Internal energy of electromagnetic or blackbody radiation.
Kinetic energy	The Kinetic energy of an object is the extra energy which it possesses due to its motion. It is defined as the work needed to accelerate a body of a given mass from rest to its current velocity. Having gained this energy during its acceleration, the body maintains this Kinetic energy unless its speed changes.

Potential energy	Potential energy can be thought of as energy stored within a physical system. It is called potential energy because it has the potential to be converted into other forms of energy, such as kinetic energy, and to do work in the process. The standard unit of measure for potential energy is the joule, the same as for work, or energy in general.
Voltage	Voltage is commonly used as a short name for electrical potential difference. In this introduction, the term 'Voltage' is however used to mean electric potential, i.e. a hypothetically measurable physical dimension, and is denoted by the algebraic variable V (inclined or italicized letter.) The SI unit for Voltage is the volt (symbol: V [not italicized].)
Work	In physics, mechanical Work is the amount of energy transferred by a force acting through a distance. Like energy, it is a scalar quantity, with SI units of joules. The term Work was first coined in the 1830s by the French mathematician Gaspard-Gustave Coriolis.
Parallel	Parallels are rectangular blocks of metal, commonly made from tool steel, stainless steel or cast iron, which have 2, 4 or 6 faces ground or lapped to a precise surface finish. They are used when machining with a miller, drill or any other machining operation that requires work to be held in a vise or with clamps - to keep work Parallel or raised evenly such as in a milling vise to give adequate height for the cutting tool/spindle to pass over. Parallels come in pairs of two, which are machined to be the same dimensions their corresponding faces.
Series circuit	If two or more circuit components are connected end to end like a daisy chain, it is said they are connected in series. A series circuit provides a single path for electric current through all of its components. If two or more circuit components are connected like the rungs of a ladder it is said they are connected in parallel.
Circuit breaker	A Circuit breaker is an automatically-operated electrical switch designed to protect an electrical circuit from damage caused by overload or short circuit. Its basic function is to detect a fault condition and, by interrupting continuity, to immediately discontinue electrical flow. Unlike a fuse, which operates once and then has to be replaced, a Circuit breaker can be reset (either manually or automatically) to resume normal operation.
Superconductivity	Superconductivity is a phenomenon occurring in certain materials generally at very low temperatures, characterized by exactly zero electrical resistance and the exclusion of the interior magnetic field. The electrical resistivity of a metallic conductor decreases gradually as the temperature is lowered. However, in ordinary conductors such as copper and silver, impurities and other defects impose a lower limit.
Watt	The Watt (symbol: W) is a derived unit of power in the International System of Units (SI.) It measures rate of energy conversion. One Watt is equivalent to 1 joule (J) of energy per second.
Second	The second (SI symbol: s), sometimes abbreviated sec., is the name of a unit of time, and is the International System of Units (SI) base unit of time. It may be measured using a clock. SI prefixes are frequently combined with the word second to denote subdivisions of the second e.g., the milli second (one thousandth of a second , the micro second (one millionth of a second , and the nano second (one billionth of a second)

Alternating current	In Alternating current (Alternating current , also ac) the movement (or flow) of electric charge periodically reverses direction. An electric charge would for instance move forward, then backward, then forward, then backward, over and over again. In direct current (DC), the movement (or flow) of electric charge is only in one direction.
Direct current	Direct current is the undirectional flow of electric charge. Direct current is produced by such sources as batteries, thermocouples, solar cells, and commutator-type electric machines of the dynamo type. Direct current may flow in a conductor such as a wire, but can also be through semiconductors, insulators, or even through a vacuum as in electron or ion beams.
Oscillation	Oscillation is the repetitive variation, typically in time, of some measure about a central value or between two or more different states. Familiar examples include a swinging pendulum and AC power. The term vibration is sometimes used more narrowly to mean a mechanical oscillation but sometimes is used to be synonymous with 'oscillation.' Oscillations occur not only in physical systems but also in biological systems and in human society.
Power station	A power station is an industrial facility for the generation of electric power. Power plant is also used to refer to the engine in ships, aircraft and other large vehicles. Some prefer to use the term energy center because it more accurately describes what the plants do, which is the conversion of other forms of energy, like chemical energy, gravitational potential energy or heat energy into electrical energy.
Transformer	A transformer is a device that transfers electrical energy from one circuit to another through inductively coupled electrical conductors. A changing current in the first circuit creates a changing magnetic field; in turn, this magnetic field induces a changing voltage in the second circuit; this is called mutual induction. By adding a load to the secondary circuit, one can make current flow in the transformer, thus transferring energy from one circuit to the other.
Mechanics	Mechanics is the branch of physics concerned with the behaviour of physical bodies when subjected to forces or displacements, and the subsequent effect of the bodies on their environment. The discipline has its roots in several ancient civilizations During the early modern period, scientists such as Galileo, Kepler, and especially Newton, laid the foundation for what is now known as classical Mechanics
Reproduction	Reproduction is the biological process by which new individual organisms are produced. Reproduction is a fundamental feature of all known life; each individual organism exists as the result of reproduction. The known methods of reproduction are broadly grouped into two main types: sexual and asexual.
Atmospheric pressure	Atmospheric pressure is sometimes defined as the force per unit area exerted against a surface by the weight of air above that surface at any given point in the Earth's atmosphere. In most circumstances Atmospheric pressure is closely approximated by the hydrostatic pressure caused by the weight of air above the measurement point. Low pressure areas have less atmospheric mass above their location, whereas high pressure areas have more atmospheric mass above their location.
Gray	Grey (international and some parts of the U.S.) or Gray describes the tints and shades ranging from black to white. These, including white and black, are known as achromatic colors or neutral colors.

Electrolysis	In chemistry and manufacturing, Electrolysis is a method of using an electric current to drive an otherwise non-spontaneous chemical reaction. Electrolysis is commercially highly important as a stage in the separation of elements from naturally-occurring sources such as ores using an electrolytic cell. · 1800 - William Nicholson and Johann Ritter decomposed water into hydrogen and oxygen.

Electricity	Electricity is a general term that encompasses a variety of phenomena resulting from the presence and flow of electric charge. These include many easily recognizable phenomena, such as lightning and static Electricity but in addition, less familiar concepts, such as the electromagnetic field and electromagnetic induction. In general usage, the word Electricity is adequate to refer to a number of physical effects.
Electromagnetism	Electromagnetism is the physics of the electromagnetic field, a field that exerts a force on particles with the property of electric charge and is reciprocally affected by the presence and motion of such particles. A changing magnetic field produces an electric field (this is the phenomenon of electromagnetic induction, the basis of operation for electrical generators, induction motors, and transformers.) Similarly, a changing electric field generates a magnetic field.
Magnetism	In physics, the term Magnetism is used to describe how materials respond on the microscopic level to an applied magnetic field; to categorize the magnetic phase of a material. For example, the most well known form of Magnetism is ferro Magnetism such that some ferromagnetic materials produce their own persistent magnetic field. Some well-known ferromagnetic materials that exhibit easily detectable magnetic properties (to form magnets) are nickel, iron, cobalt, gadolinium and their alloys.
Sensor	A sensor is a device that measures a physical quantity and converts it into a signal which can be read by an observer or by an instrument. For example, a mercury thermometer converts the measured temperature into expansion and contraction of a liquid which can be read on a calibrated glass tube. A thermocouple converts temperature to an output voltage which can be read by a voltmeter.
Wave	A Wave is a disturbance that propagates through space and time, usually with transference of energy. A mechanical Wave is a Wave that propagates or travels through a medium due to the restoring forces it produces upon deformation. There also exist Wave s capable of traveling through a vacuum, including electromagnetic radiation and probably gravitational radiation.
Waves	Waves is a three-part novel by Ogan Gurel published in 2009. A 21st century version of Faust, the novel explores good and evil in both individual and global settings focusing around a hypothetical technology that has both medical and military applications. The protagonist, Tomas Twarok, is a contemplative and idealistic doctor-turned-entrepreneur who strikes a deal with his college friend, Maximilian Iblis, a ruthless hedge fund manager.
Magnet	A Magnet is a material or object that produces a Magnet ic field. This Magnet ic field is invisible but is responsible for the most notable property of a Magnet a force that pulls on other ferro Magnet ic materials and attracts or repels other Magnet s. Some materials can be Magnet ised by a Magnet ic field but do not remain Magnet ised when the field is removed; these are termed soft.
Magnetic field	Magnetic field s surround magnetic materials and electric currents and are detected by the force they exert on other magnetic materials and moving electric charges. The Magnetic field at any given point is specified by both a direction and a magnitude (or strength); as such it is a vector field. For the physics of magnetic materials, see magnetism and magnet, more specifically ferromagnetism, paramagnetism, and diamagnetism.

North pole	The North Pole is, subject to the caveats explained below, defined as the point in the northern hemisphere where the Earth's axis of rotation meets the Earth's surface. It should not be confused with the North Magnetic Pole. The North Pole is the northernmost point on Earth, lying diametrically opposite the South Pole.
South pole	The South Pole is the southernmost point on the surface of the Earth. It lies on the continent of Antarctica, on the opposite side of the Earth from the North Pole. It is the site of the United States Amundsen-Scott South Pole Station, which was established in 1956 and has been permanently staffed since that year.
Earth	Earth is the third planet from the Sun. It is the fifth largest of the eight planets in the solar system, and the largest of the terrestrial planets (non-gas planets) in the Solar System in terms of diameter, mass and density. It is also referred to as the World, the Blue Planet, and Terra.
Magnetic declination	The Magnetic declination at any point on the Earth is the angle between the local magnetic field--the direction the north end of a compass points--and true north. The declination is positive when the magnetic north is east of true north. The term magnetic variation is equivalent, and is more often used in aeronautical and other forms of navigation.
Baculometry	Bacculometry is the art of measuring accessible or inaccessible distances by the help of one or more baculi, staves in his Elements. Jacques Ozanam also gives an illustration of the principles of Baculometry
Electric field	In physics, the space surrounding an electric charge or in the presence of a time-varying magnetic field has a property called an Electric field This Electric field exerts a force on other electrically charged objects. The concept of an Electric field was introduced by Michael Faraday.
Electrostatic	Electrostatic s is the branch of science that deals with the phenomena arising from stationary or slowly moving electric charges. Since classical antiquity it was known that some materials such as amber attract light particles after rubbing. The Greek word for amber, Î®λεκτρον , was the source of the word 'electricity'.
Elementary particle	In particle physics, an Elementary particle or fundamental particle is a particle not known to have substructure; that is, it is not known to be made up of smaller particles. If an Elementary particle truly has no substructure, then it is one of the basic building blocks of the universe from which all other particles are made. In the Standard Model, the quarks, leptons, and gauge bosons are Elementary particle s.
Magnetic monopole	In physics, a Magnetic monopole is a hypothetical particle that is a magnet with only one pole In more technical terms, it would have a net 'magnetic charge'. Modern interest in the concept stems from particle theories, notably Grand Unified Theories and superstring theories, which predict their existence.
Meissner effect	The Meissner effect is the expulsion of a magnetic field from a superconductor. Walther Meissner and Robert Ochsenfeld discovered the phenomenon in 1933 by measuring the magnetic field distribution outside tin and lead samples. The samples, in the presence of an applied magnetic field, were cooled below what is called their superconducting transition temperature.
Positive	In astrology, a positive dominant, active, yang, diurnal or masculine sign refers to any of the six odd-numbered signs of the zodiac: Aries, Gemini, Leo, Libra, Sagittarius or Aquarius.

These signs constitute the fire and air triplicities. They are termed positive because they tend to be more communicative, buoyant and sociable - broadly matching the extrovert personality, aiming for a large number of friends and to focusing on working with them rather than on their own resources.

Subatomic

A subatomic particle is an elementary or composite particle smaller than an atom. Particle physics and nuclear physics are concerned with the study of these particles, their interactions, and non-atomic matter.

subatomic particles include the atomic constituents electrons, protons, and neutrons.

Subatomic particle

A subatomic particle is an elementary or composite particle smaller than an atom. Particle physics and nuclear physics are concerned with the study of these particles, their interactions, and non-atomic matter.

Subatomic particles include the atomic constituents electrons, protons, and neutrons.

Subatomic particles

In physics, subatomic particles are the particles composing nucleons and atoms. There are two types of subatomic particles elementary particles, which are not made of other particles, and composite particles. Particle physics and nuclear physics study these particles and how they interact.

Superconductivity

Superconductivity is a phenomenon occurring in certain materials generally at very low temperatures, characterized by exactly zero electrical resistance and the exclusion of the interior magnetic field.

The electrical resistivity of a metallic conductor decreases gradually as the temperature is lowered. However, in ordinary conductors such as copper and silver, impurities and other defects impose a lower limit.

Electromagnet

An Electromagnet is a type of magnet in which the magnetic field is produced by the flow of electric current. The magnetic field disappears when the current ceases. Electromagnet attracts paper clips when current is applied creating a magnetic field, loses them when current and magnetic field are removed

A wire with an electric current passing through it, generates a magnetic field around it, this is a simple Electromagnet

Observation

Observation is either an activity of a living being (such as a human), consisting of receiving knowledge of the outside world through the senses, or the recording of data using scientific instruments. The term may also refer to any datum collected during this activity.

The scientific method requires Observation s of nature to formulate and test hypotheses.

Alternating current

In Alternating current (Alternating current , also ac) the movement (or flow) of electric charge periodically reverses direction. An electric charge would for instance move forward, then backward, then forward, then backward, over and over again. In direct current (DC), the movement (or flow) of electric charge is only in one direction.

Electric motor

An Electric motor is a device using electrical energy to produce mechanical energy, nearly always by the interaction of magnetic fields and current-carrying conductors. The reverse process, that of using mechanical energy to produce electrical energy, is accomplished by a generator or dynamo. Traction motors used on vehicles often perform both tasks.

Electron

An Electron is a subatomic particle that carries a negative electric charge. It has no known substructure and is believed to be a point particle. An Electron has a mass that is approximately 1836 times less than that of the proton.

Force

In physics, a Force is any external agent that causes a change in the motion of a free body, or that causes stress in a fixed body. It can also be described by intuitive concepts such as a push or pull that can cause an object with mass to change its velocity , i.e., to accelerate, or which can cause a flexible object to deform. Force has both magnitude and direction, making it a vector quantity.

Speaker

A loudspeaker, speaker, or speaker system is an electroacoustical transducer that converts an electrical signal to sound. The term loudspeaker can refer to individual transducers, or to complete systems consisting of a enclosure incorporating one or more drivers and electrical filter components. Loudspeakers, just as with other electroacoustic transducers, are the most variable elements in an audio system and are responsible for the greatest degree of audible differences between sound systems.

Nuclear fusion

In physics and nuclear chemistry, nuclear fusion is the process by which multiple like-charged atomic nuclei join together to form a heavier nucleus. It is accompanied by the release or absorption of energy. Iron and nickel nuclei have the largest binding energies per nucleon of all nuclei.

Plasma

In physics and chemistry, plasma is an ionized gas, in which a certain proportion of electrons are free rather than being bound to an atom or molecule. The ability of the positive and negative charges to move somewhat independently makes the plasma electrically conductive so that it responds strongly to electromagnetic fields. Plasma therefore has properties quite unlike those of solids, liquids or gases and is considered to be a distinct state of matter.

Sodium

Sodium is an element which has the symbol N, atomic number 11, atomic mass 22.9898 g/mol, common oxidation number +1. Sodium is a soft, silvery white, highly reactive element and is a member of the alkali metals within 'group 1'. It has only one stable isotope, ^{23}Na.

Electromagnetic induction

Electromagnetic induction is the production of voltage across a conductor situated in a changing magnetic field or a conductor moving through a stationary magnetic field.

Michael Faraday is generally credited with having discovered the induction phenomenon in 1831 though it may have been anticipated by the work of Francesco Zantedeschi in 1829. Around 1830 to 1832 Joseph Henry made a similar discovery, but did not publish his findings until later.

Particle accelerator

A particle accelerator is a device that uses electric fields to propel electrically-charged particles to high speeds and to contain them. An ordinary CRT television set is a simple form of accelerator. There are two basic types: linear accelerators and circular accelerators.

Tevatron

Tevatron is a circular particle accelerator at the Fermi National Accelerator Laboratory in Batavia, Illinois and is the highest energy particle collider in the world until the collisions begin at the Large Hadron Collider. The Tevatron is a synchrotron that accelerates protons and antiprotons in a 6.3 km ring to energies of up to 1 TeV, hence the name. The Tevatron was completed in 1983 at a cost of $120 million and has been regularly upgraded since then.

Mechanical energy

In physics, Mechanical energy describes the sum of potential energy and kinetic energy present in the components of a mechanical system.

Scientists make simplifying assumptions to make calculations about how mechanical systems react. For example, instead of calculating the Mechanical energy separately for each of the billions of molecules in a soccer ball, it is easier to treat the entire ball as one object.

Power station

A power station is an industrial facility for the generation of electric power.

Power plant is also used to refer to the engine in ships, aircraft and other large vehicles. Some prefer to use the term energy center because it more accurately describes what the plants do, which is the conversion of other forms of energy, like chemical energy, gravitational potential energy or heat energy into electrical energy.

Direct current

Direct current is the undirectional flow of electric charge. Direct current is produced by such sources as batteries, thermocouples, solar cells, and commutator-type electric machines of the dynamo type. Direct current may flow in a conductor such as a wire, but can also be through semiconductors, insulators, or even through a vacuum as in electron or ion beams.

Spark

In mathematics, specifically in linear algebra, the Spark of a matrix A is the smallest number n such that there exists a set of n columns in A which are linearly dependent. Formally,

$$\text{spark}(A) = \min_{d \neq 0} \|d\|_0 \text{ s.t. } Ad = 0.$$

By contrast, the rank of a matrix is the smallest number k such that all sets of k + 1 columns in A are linearly dependent.

The concept of the Spark is of use in the theory of compressive sensing, where requirements on the Spark of the measurement matrix are used to ensure stability and consistency of various estimation techniques.

Transformer

A transformer is a device that transfers electrical energy from one circuit to another through inductively coupled electrical conductors. A changing current in the first circuit creates a changing magnetic field; in turn, this magnetic field induces a changing voltage in the second circuit; this is called mutual induction. By adding a load to the secondary circuit, one can make current flow in the transformer, thus transferring energy from one circuit to the other.

Voltage

Voltage is commonly used as a short name for electrical potential difference. In this introduction, the term 'Voltage' is however used to mean electric potential, i.e. a hypothetically measurable physical dimension, and is denoted by the algebraic variable V (inclined or italicized letter.) The SI unit for Voltage is the volt (symbol: V [not italicized].)

Mechanics

Mechanics is the branch of physics concerned with the behaviour of physical bodies when subjected to forces or displacements, and the subsequent effect of the bodies on their environment. The discipline has its roots in several ancient civilizations During the early modern period, scientists such as Galileo, Kepler, and especially Newton, laid the foundation for what is now known as classical Mechanics

Reproduction

Reproduction is the biological process by which new individual organisms are produced. Reproduction is a fundamental feature of all known life; each individual organism exists as the result of reproduction. The known methods of reproduction are broadly grouped into two main types: sexual and asexual.

Transducer

A transducer is a device, usually electrical, electronic, electro-mechanical, electromagnetic, photonic, or photovoltaic that converts one type of energy or physical attribute to another for various purposes including measurement or information transfer.

The term transducer is commonly used in two senses; the sensor, used to detect a parameter in one form and report it in another, and the audio loudspeaker, which converts electrical voltage variations representing music or speech, to mechanical cone vibration and hence vibrates air molecules creating acoustical energy.

· Electromagnetic:

· Antenna - converts electromagnetic waves into electric current and vice versa.
· Cathode ray tube - converts electrical signals into visual form
· Fluorescent lamp, light bulb - converts electrical power into visible light
· Magnetic cartridge - converts motion into electrical form
· Photodetector or Photoresistor - converts changes in light levels into resistance changes
· Tape head - converts changing magnetic fields into electrical form
· Hall effect sensor - converts a magnetic field level into electrical form only.
· Electrochemical:

· pH probes
· Electro-galvanic fuel cell
· Electromechanical:

· Electroactive polymers
· Galvanometer
· MEMS
· Rotary motor, linear motor
· Vibration powered generator
· Potentiometer when used for measuring position
· Load cell converts force to mV/V electrical signal using strain gauge
· Accelerometer
· Strain gauge
· String Potentiometer
· Air flow sensor
· Electroacoustic:

· Geophone - convert a ground movement into voltage
· Gramophone pick-up
· Hydrophone - converts changes in water pressure into an electrical form
· Loudspeaker, earphone - converts changes in electrical signals into acoustic form
· Microphone - converts changes in air pressure into an electrical signal
· Piezoelectric crystal - converts pressure changes into electrical form
· Tactile transducer
· Photoelectric:

· Laser diode, light-emitting diode - convert electrical power into forms of light
· Photodiode, photoresistor, phototransistor, photomultiplier tube - converts changing light levels into electrical form
· Electrostatic:

· Electrometer
· Thermoelectric:

· RTD Resistance Temperature Detector
· Thermocouple
· Peltier cooler
· Thermistor
· Radioacoustic:

· Geiger-Müller tube used for measuring radioactivity.
· Receiver .

VCR

The videocassette recorder (or VCR more commonly known in the UK and Ireland as the video recorder), is a type of video tape recorder that uses removable videotape cassettes containing magnetic tape to record audio and video from a television broadcast so it can be played back later. Most VCR s have their own tuner (for direct TV reception) and a programmable timer (for unattended recording of a certain channel at a particular time.) A VHS VCR manufactured by Metz. The insides of a VCR

The history of the videocassette recorder follows the history of videotape recording in general.

Tap

A transformer Tap is a connection point along a transformer winding that allows a certain number of turns to be selected. By this means, a transformer with a variable turns ratio is produced, enabling voltage regulation of the output. The Tap selection is made via a Tap changer mechanism.

Frequency

Frequency is the number of occurrences of a repeating event per unit time. It is also referred to as temporal Frequency The period is the duration of one cycle in a repeating event, so the period is the reciprocal of the Frequency

Waveform

Waveform means the shape and form of a signal such as a wave moving in a solid, liquid or gaseous medium.

In many cases the medium in which the wave is being propagated does not permit a direct visual image of the form. In these cases, the term 'waveform' refers to the shape of a graph of the varying quantity against time or distance.

Oscillation

Oscillation is the repetitive variation, typically in time, of some measure about a central value or between two or more different states. Familiar examples include a swinging pendulum and AC power. The term vibration is sometimes used more narrowly to mean a mechanical oscillation but sometimes is used to be synonymous with 'oscillation.' Oscillations occur not only in physical systems but also in biological systems and in human society.

Transverse wave

A transverse wave is a moving wave that consists of oscillations occurring perpendicular to the direction of energy transfer. If a transverse wave is moving in the positive x-direction, its oscillations are in up and down directions that lie in the yz-plane. A transverse wave could be represented by moving a ribbon or piece of string, spread across a table, to the left and right or up and down.

Amplitude

Amplitude is the magnitude of change in the oscillating variable, with each oscillation, within an oscillating system. For instance, sound waves are oscillations in atmospheric pressure and their Amplitude s are proportional to the change in pressure during one oscillation. If the variable undergoes regular oscillations, and a graph of the system is drawn with the oscillating variable as the vertical axis and time as the horizontal axis, the Amplitude is visually represented by the vertical distance between the extrema of the curve.

Vacuum

A vacuum is a volume of space that is essentially empty of matter, such that its gaseous pressure is much less than atmospheric pressure. The word comes from the Latin term for 'empty,' but in reality, no volume of space can ever be perfectly empty. A perfect vacuum with a gaseous pressure of absolute zero is a philosophical concept that is never observed in practice.

Wavelength

In physics wavelength is the distance between repeating units of a propagating wave of a given frequency. It is commonly designated by the Greek letter lambda. Examples of wave-like phenomena are light, water waves, and sound waves.

X-radiation

X-radiation is a form of electromagnetic radiation. X-rays have a wavelength in the range of 10 to 0.01 nanometers, corresponding to frequencies in the range 30 petahertz to 30 exahertz and energies in the range 120 eV to 120 keV. They are longer than gamma rays but shorter than UV rays.

Atom

The Atom is a basic unit of matter consisting of a dense, central nucleus surrounded by a cloud of negatively charged electrons. The Atom ic nucleus contains a mix of positively charged protons and electrically neutral neutrons (except in the case of hydrogen-1, which is the only stable nuclide with no neutron.) The electrons of an Atom are bound to the nucleus by the electromagnetic force.

Big Bang

The Big Bang is the cosmological model of the initial conditions and subsequent development of the universe that is supported by the most comprehensive and accurate explanations from current scientific evidence and observation. As used by cosmologists, the term Big Bang generally refers to the idea that the universe has expanded from a primordial hot and dense initial condition at some finite time in the past (currently estimated to have been approximately 13.7 billion years ago), and continues to expand to this day.

Georges Lemaître proposed what became known as the Big Bang theory of the origin of the Universe, although he called it his 'hypothesis of the primeval atom'.

Energy

In physics, Energy is a scalar physical quantity that describes the amount of work that can be performed by a force, an attribute of objects and systems that is subject to a conservation law. Different forms of Energy include kinetic, potential, thermal, gravitational, sound, light, elastic, and electromagnetic Energy The forms of Energy are often named after a related force.

Gamma rays

Gamma rays (denoted as γ) are electromagnetic radiation of high energy. They are produced by sub-atomic particle interactions, such as electron-positron annihilation, neutral pion decay, radioactive decay, fusion, fission or inverse Compton scattering in astrophysical processes. Gamma rays typically have frequencies above 10^{19} Hz and therefore energies above 100 keV and wavelength less than 10 picometers, often smaller than an atom.

Heat

In physics and thermodynamics, Heat is the process of energy transfer from one body or system due to thermal contact, which in turn is defined as an energy transfer to a body in any other way than due to work performed on the body.

When an infinitesimal amount of Heat δQ is tranferred to a body in thermal equilibrium at absolute temperature T in a reversible way, then it is given by the quantity TdS, where S is the entropy of the body.

A related term is thermal energy, loosely defined as the energy of a body that increases with its temperature.

Matter

The term Matter traditionally refers to the substance that objects are made of. One common way to identify this 'substance' is through its physical properties; a common definition of Matter is anything that has mass and occupies a volume. However, this definition has to be revised in light of quantum mechanics, where the concept of 'having mass', and 'occupying space' are not as well-defined as in everyday life.

Molecule

A Molecule is defined as a sufficiently stable, electrically neutral group of at least two atoms in a definite arrangement held together by very strong (covalent) chemical bonds. Molecule s are distinguished from polyatomic ions in this strict sense. In organic chemistry and biochemistry, the term Molecule is used less strictly and also is applied to charged organic Molecule s and bio Molecule s.

Nucleus

The nucleus of an atom is the very dense region, consisting of nucleons, at the center of an atom. Although the size of the nucleus varies considerably according to the mass of the atom, the size of the entire atom is comparatively constant. Almost all of the mass in an atom is made up from the protons and neutrons in the nucleus with a very small contribution from the orbiting electrons.

Proton

The Proton is a subatomic particle with an electric charge of +1 elementary charge. It is found in the nucleus of each atom but is also stable by itself and has a second identity as the hydrogen ion, $^{1}H^{+}$. It is composed of three even more fundamental particles comprising two up quarks and one down quark.

Radiation

Radiation, as in physics, is energy in the form of waves or moving subatomic particles emitted by an atom or other body as it changes from a higher energy state to a lower energy state. Radiation can be classified as ionizing or non-ionizing radiation, depending on its effect on atomic matter. The most common use of the word 'radiation' refers to ionizing radiation.

Radio

Radio is the transmission of signals, by modulation of electromagnetic waves with frequencies below those of visible light. Electromagnetic radiation travels by means of oscillating electromagnetic fields that pass through the air and the vacuum of space. Information is carried by systematically changing some property of the radiated waves, such as amplitude, frequency, or phase.

Radio waves

Radio waves are electromagnetic waves occurring on the radio frequency portion of the electromagnetic spectrum. A common use is to transport information through the atmosphere or outer space without wires. Radio waves are distinguished from other kinds of electromagnetic waves by their wavelength, a relatively long wavelength in the electromagnetic spectrum.

Transmitter

A transmitter is an electronic device which, usually with the aid of an antenna, propagates an electromagnetic signal such as radio, television, or other telecommunications. In other applications signals can also be transmitted using an analog 0/4-20 mA current loop signal. WDET-FM transmitter

Generally and in communication and information processing, a transmitter is any object which sends information to an observer.

Ultraviolet	Ultraviolet light is electromagnetic radiation with a wavelength shorter than that of visible light, but longer than X-rays. It is so named because the spectrum consists of electromagnetic waves with frequencies higher than those that humans identify as the color violet. UV light is typically found as part of the radiation received by the Earth from the Sun.
Visible light	The visible spectrum is the portion of the electromagnetic spectrum that is visible to the human eye. Electromagnetic radiation in this range of wavelengths is called visible light or simply light. A typical human eye will respond to wavelengths in air from about 380 to 750 nm.
Atmosphere	An Atmosphere is a layer of gases that may surround a material body of sufficient mass, by the gravity of the body, and are retained for a longer duration if gravity is high and the Atmosphere s temperature is low. Some planets consist mainly of various gases, but only their outer layer is their Atmosphere . The term stellar Atmosphere describes the outer region of a star, and typically includes the portion starting from the opaque photosphere outwards.
Collision	A Collision is an isolated event in which two or more moving bodies (colliding bodies) exert relatively strong forces on each other for a relatively short time. Deflection happens when an object hits a plane surface Collision s involve forces (there is a change in velocity.) Collision s can be elastic, meaning they conserve energy and momentum, inelastic, meaning they conserve momentum but not energy, or totally inelastic (or plastic), meaning they conserve momentum and the two objects stick together.
Electric charge	Electric charge is a fundamental conserved property of some subatomic particles, which determines their electromagnetic interaction. Electrically charged matter is influenced by, and produces, electromagnetic fields. The interaction between a moving charge and an electromagnetic field is the source of the electromagnetic force, which is one of the four fundamental forces.
Electromagnetic force	In physics, the Electromagnetic force is the force that the electromagnetic field exerts on electrically charged particles. It is the Electromagnetic force that holds electrons and protons together in atoms, and which hold atoms together to make molecules. The Electromagnetic force operates via the exchange of messenger particles called photons and virtual photons.
Reflection	Reflection is the change in direction of a wave front at an interface between two different media so that the wave front returns into the medium from which it originated. Common examples include the reflection of light, sound and water waves. Law of reflection: Angle of incidence = Angle of reflection Reflections may occur in a number of wave and particle phenomena; these include acoustic, seismic waves in geologic structures, surface waves in bodies of water, and various electromagnetic waves, most usefully from VHF and higher radar frequencies, progressing upward through centimeter to millimeter-wavelength radar and the various light frequencies and (with special 'grazing' mirrors, to X-ray frequencies and beyond to gamma rays.

Spectrum	A spectrum is a condition that is not limited to a specific set of values but can vary infinitely within a continuum. The word saw its first scientific use within the field of optics to describe the rainbow of colors in visible light when separated using a prism; it has since been applied by analogy to many fields other than optics. Thus, one might talk about the spectrum of political opinion, or the spectrum of activity of a drug, or the autism spectrum.
Convection	Convection in the most general terms refers to the movement of molecules within fluids (i.e. liquids, gases and rheids.) Convection is one of the major modes of heat transfer and mass transfer. In fluids, convective heat and mass transfer take place through both diffusion - the random Brownian motion of individual particles in the fluid - and by advection, in which matter or heat is transported by the larger-scale motion of currents in the fluid.
Polarity	In chemistry, polarity refers to the dipole-dipole intermolecular forces between the slightly positively-charged end of one molecule to the negative end of another or the same molecule. Molecular polarity is dependent on the difference in electronegativity between atoms in a compound and the asymmetry of the compound's structure. For example, water is thought to be polar because of the unequal sharing of its electrons.
Radar	Radar is a system that uses electromagnetic waves to identify the range, altitude, direction, or speed of both moving and fixed objects such as aircraft, ships, motor vehicles, weather formations, and terrain. The term RADAR was coined in 1941 as an acronym for Radio Detection and Ranging. The term has since entered the English language as a standard word, radar, losing the capitalization.
Satellite	In the context of spaceflight, a satellite is an object which has been placed into orbit by human endeavor. Such objects are sometimes called artificial satellite s to distinguish them from natural satellite s such as the Moon. The first artificial satellite Sputnik 1, was launched by the Soviet Union in 1957.
Space	Space is the boundless, three-dimensional extent in which objects and events occur and have relative position and direction. Physical Space is often conceived in three linear dimensions, although modern physicists usually consider it, with time, to be part of the boundless four-dimensional continuum known as Space time. In mathematics Space s with different numbers of dimensions and with different underlying structures can be examined.
Venus	Venus is the second-closest planet to the Sun, orbiting it every 224.7 Earth days. The planet is named after Venus, the Roman goddess of love. It is the brightest natural object in the night sky, except for the Moon, reaching an apparent magnitude of -4.6.
Personal digital assistant	A personal digital assistant is a handheld computer also known as small or palmtop computers. Newer PDAs also have both color screens and audio capabilities, enabling them to be used as mobile phones,, web browsers, or portable media players. Many PDAs can access the Internet, intranets or extranets via Wi-Fi, or Wireless Wide-Area Networks.
Wireless communication	Wireless communication is the transfer of information over a distance without the use of electrical conductors or 'wires'. The distances involved may be short or very long. When the context is clear the term is often simply shortened to 'wireless'.

Ozone

Ozone or trioxygen is a triatomic molecule, consisting of three oxygen atoms. It is an allotrope of oxygen that is much less stable than the diatomic O_2. Ground-level ozone is an air pollutant with harmful effects on the respiratory systems of animals and humans.

Ozone layer

The photochemical mechanisms that give rise to the ozone layer were worked out by the British physicist Sidney Chapman in 1930. Ozone in the earth's stratosphere is created by ultraviolet light striking oxygen molecules containing two oxygen atoms, splitting them into individual oxygen atoms; surface. About 90% of the ozone in our atmosphere is contained in the stratosphere.

N rays

N rays (or N-rays) are a hypothesized form of radiation, described by French scientist René-Prosper Blondlot, and initially confirmed by others, but subsequently found to be illusory.

In 1903, Blondlot, a distinguished physicist who was one of 8 physicists who were corresponding members of the French Academy of Science announced his discovery while working at the University of Nancy attempting to polarize X-rays. He had perceived changes in the brightness of an electric spark in a spark gap placed in an X-ray beam which he photographed and he later attributed to the novel form of radiation, naming it the N-ray for the University of Nancy.

X-rays

X-radiation is a form of electromagnetic radiation. X-rays have a wavelength in the range of 10 to 0.01 nanometers, corresponding to frequencies in the range 30 petahertz to 30 exahertz and energies in the range 120 eV to 120 keV. They are longer than gamma rays but shorter than UV rays.

Periodic table

The periodic table of the chemical elements is a tabular method of displaying the chemical elements. Although precursors to this table exist, its invention is generally credited to Russian chemist Dmitri Mendeleev in 1869. Mendeleev intended the table to illustrate recurring trends in the properties of the elements.

Physicist

A Physicist is a scientist who studies or practices physics. Physicist s study a wide range of physical phenomena in many branches of physics spanning all length scales: from sub-atomic particles of which all ordinary matter is made (particle physics) to the behavior of the material Universe as a whole (cosmology.)

Most material a student encounters in the undergraduate physics curriculum is based on discoveries and insights of a century or more in the past.

Radiant

The radiant of a meteor shower is the point in the sky that meteors appear to originate from. For example, the Perseids are meteors which appear to come from a point within the constellation of Perseun.

Radiant energy

Radiant energy is the energy of electromagnetic waves. The quantity of Radiant energy may be calculated by integrating radiant flux (or power) with respect to time and, like all forms of energy, its SI unit is the joule. The term is used particularly when radiation is emitted by a source into the surrounding environment.

Temperature

Temperature is a physical property of a system that underlies the common notions of hot and cold; something that is hotter generally has the greater temperature. Specifically, temperature is a property of matter. Temperature is one of the principal parameters of thermodynamics.

Star	A star is a massive, luminous ball of plasma that is held together by gravity. The nearest star to Earth is the Sun, which is the source of most of the energy on Earth. Other star s are visible in the night sky, when they are not outshone by the Sun.
Pyrometer	Pyrometer is any non-contacting device that intercepts and measures thermal radiation. This measure is used to determine temperature, often of the object's surface. The word pyrometer comes from the Greek word for fire, 'πυρ', and meter, meaning to measure.
Thermometer	A Thermometer is a device that measures temperature or temperature gradient using a variety of different principles. A Thermometer has two important elements: the temperature sensor (e.g. the bulb on a mercury Thermometer in which some physical change occurs with temperature, plus some means of converting this physical change into a value (e.g. the scale on a mercury Thermometer) Thermometer s increasingly use electronic means to provide a digital display or input to a computer.
Noise	Noise in audio, recording, and broadcast systems refers to the residual low level sound (usually hiss and hum) that is heard in quiet periods of a programme. In audio engineering it can refer either to the acoustic Noise from loudspeakers, or to the unwanted residual electronic Noise signal that gives rise to acoustic Noise heard as 'hiss'. This signal Noise is commonly measured using A-weighting or ITU-R 468 weighting Noise is often generated deliberately and used as a test signal.
Thermodynamic	In physics, Thermodynamic s '>power') is the study of the conversion of energy into work and heat and its relation to macroscopic variables such as temperature,volume and pressure. Its underpinnings, based upon statistical predictions of the collective motion of particles from their microscopic behavior, is the field of statistical Thermodynamic s, a branch of statistical mechanics. Historically, Thermodynamic s developed out of need to increase the efficiency of early steam engines.
Thermodynamics	In physics, thermodynamics is the study of the transformation of energy into different forms and its relation to macroscopic variables such as temperature, pressure, and volume. Its underpinnings, based upon statistical predictions of the collective motion of particles from their microscopic behavior, is the field of statistical thermodynamics, a branch of statistical mechanics. Roughly, heat means 'energy in transit' and dynamics relates to 'movement'; thus, in essence thermodynamics studies the movement of energy and how energy instills movement.
Universe	The Universe is defined as everything that physically exists: the entirety of space and time, all forms of matter, energy and momentum, and the physical laws and constants that govern them. However, the term 'universe' may be used in slightly different contextual senses, denoting such concepts as the cosmos, the world or Nature. Astronomical observations indicate that the universe is 13.73 ± 0.12 billion years old and at least 93 billion light years across.
Wilkinson Microwave Anisotropy Probe	The Wilkinson Microwave Anisotropy Probe -- also known as the Microwave Anisotropy Probe, and Explorer 80 -- measures the temperature of the Big Bang's remnant radiant heat. Headed by Professor Charles L. Bennett, Johns Hopkins University, the mission is a joint project between the NASA Goddard Space Flight Center and Princeton University.

Air	The Earth's atmosphere is a layer of gases surrounding the planet Earth that is retained by Earth's gravity. The atmosphere protects life on Earth by absorbing ultraviolet solar radiation, warming the surface through heat retention (greenhouse effect), and reducing temperature extremes between day and night. Dry Air contains roughly (by volume) 78.08% nitrogen, 20.95% oxygen, 0.93% argon, 0.038% carbon dioxide, and trace amounts of other gases.
Polar stratospheric clouds	Polar stratospheric clouds are clouds in the winter polar stratosphere at altitudes of 15,000-25,000 metres. They are implicated in the formation of ozone holes; their effects on ozone depletion arise because they support chemical reactions that produce active chlorine which catalyzes ozone destruction, and also because they remove gaseous nitric acid, perturbing nitrogen and chlorine cycles in a way which increases ozone destruction. The stratosphere is very dry; unlike the troposphere, it rarely allows clouds to form.
Refrigerator	A Refrigerator (often called a 'fridge' for short) is a cooling appliance comprising a thermally insulated compartment and a heat pump--chemical or mechanical means--to transfer heat from it to the external environment, cooling the contents to a temperature below ambient. Refrigerator s are extensively used to store foods which spoil from bacterial growth if not refrigerated. A device described as a Refrigerator maintains a temperature a few degrees above the freezing point of water; a similar device which maintains a temperature below the freezing point of water is called a 'freezer.' The Refrigerator is a relatively modern invention among kitchen appliances.
Glass	Glass generally refers to hard, brittle, transparent material, such as those used for windows, many bottles, or eyewear. Examples of such solid materials include, but are not limited to, soda-lime Glass borosilicate Glass acrylic Glass sugar Glass Muscovy Glass or aluminium oxynitride. In the technical sense, Glass is an inorganic product of fusion which has been cooled through the Glass transition to a rigid condition without crystallizing.
Astronomy	Astronomy , 'law') is the scientific study of celestial objects (such as stars, planets, comets, and galaxies) and phenomena that originate outside the Earth's atmosphere (such as the cosmic background radiation.) It is concerned with the evolution, physics, chemistry, meteorology, and motion of celestial objects, as well as the formation and development of the universe.
Evaporation	Evaporation is the slow vaporization of a liquid and the reverse of condensation. A type of phase transition, it is the process by which molecules in a liquid state (e.g. water) spontaneously become gaseous (e.g. water vapor.) Generally, Evaporation can be seen by the gradual disappearance of a liquid from a substance when exposed to a significant volume of gas.
Humidity	Humidity is the amount of water vapour in the air. In daily language the term Humidity is normally taken to mean relative Humidity Relative Humidity is defined as the ratio of the partial pressure of water vapour in a parcel of air to the saturated vapour pressure of water vapour at a prescribed temperature.
Ionosphere	The Ionosphere is the uppermost part of the atmosphere, distinguished because it is ionized by solar radiation. It plays an important part in atmospheric electricity and forms the inner edge of the magnetosphere. It has practical importance because, among other functions, it influences radio propagation to distant places on the Earth.

Ocean	An ocean is a major body of saline water, and a principal component of the hydrosphere. Approximately 71% of the Earth's surface (an area of some 361 million square kilometers) is covered by ocean a continuous body of water that is customarily divided into several principal ocean s and smaller seas. More than half of this area is over 3,000 meters (9,800 ft) deep.
Spacecraft	A spacecraft is a vehicle or machine designed for spaceflight. On a sub-orbital spaceflight, a spacecraft enters outer space but then returns to the planetary surface without making a complete orbit. For an orbital spaceflight, a spacecraft enters a closed orbit around the planetary body.
Telescope	A telescope is an instrument designed for the observation of remote objects by the collection of electromagnetic radiation. The first known practically functioning telescope s were invented in the Netherlands at the beginning of the 17th century. ' telescope s' can refer to a whole range of instruments operating in most regions of the electromagnetic spectrum.
Water vapor	Water vapor or water vapour, also aqueous vapor, is the gas phase of water. Water vapor is one state of the water cycle within the hydrosphere. Water vapor can be produced from the evaporation of liquid water or from the sublimation of ice.
Oersted	Oersted is the unit of magnetizing field in the CGS system of units. It is defined as $1000/4\pi$ amperes per meter of flux path, in terms of SI units. The oersted is closely related to the gauss, the CGS unit of magnetic field.
Electrolysis	In chemistry and manufacturing, Electrolysis is a method of using an electric current to drive an otherwise non-spontaneous chemical reaction. Electrolysis is commercially highly important as a stage in the separation of elements from naturally-occurring sources such as ores using an electrolytic cell. · 1800 - William Nicholson and Johann Ritter decomposed water into hydrogen and oxygen.
Telegraph	Telegraphy is the long-distance transmission of written messages without physical transport of letters. Radiotelegraphy or wireless telegraphy transmits messages using radio. Telegraphy includes recent forms of data transmission such as fax, email, and computer networks in general. A telegraph is a machine for transmitting and receiving messages over long distances.

Mirror	A Mirror or looking glass is an object with at least one polished and therefore specularly reflective surface. The most familiar type of Mirror is the plane Mirror which has a flat surface. Curved Mirror s are also used, to produce magnified or diminished images or focus light or simply distort the reflected image.
Optics	Optics is the science that describes the behavior and properties of light and the interaction of light with matter. Optics explains optical phenomena. The word optics comes from á½€πτικÎ®, meaning appearance or look in Ancient Greek.)
Transverse wave	A transverse wave is a moving wave that consists of oscillations occurring perpendicular to the direction of energy transfer. If a transverse wave is moving in the positive x-direction, its oscillations are in up and down directions that lie in the yz-plane. A transverse wave could be represented by moving a ribbon or piece of string, spread across a table, to the left and right or up and down.
Diffuse reflection	Diffuse reflection is the reflection of light from an uneven or granular surface such that an incident ray is seemingly reflected at a number of angles. It is the complement to specular reflection. If a surface is completely nonspecular, the reflected light will be evenly spread over the hemisphere surrounding the surface (2π steradians.)
Reflection	Reflection is the change in direction of a wave front at an interface between two different media so that the wave front returns into the medium from which it originated. Common examples include the reflection of light, sound and water waves. Law of reflection: Angle of incidence = Angle of reflection Reflections may occur in a number of wave and particle phenomena; these include acoustic, seismic waves in geologic structures, surface waves in bodies of water, and various electromagnetic waves, most usefully from VHF and higher radar frequencies, progressing upward through centimeter to millimeter-wavelength radar and the various light frequencies and (with special 'grazing' mirrors, to X-ray frequencies and beyond to gamma rays.
Specular reflection	Specular reflection is the perfect, mirror-like reflection of light from a surface, in which light from a single incoming direction is reflected into a single outgoing direction. Such behavior is described by the law of reflection, which states that the direction of incoming light, and the direction of outgoing light reflected make the same angle with respect to the surface normal, thus the angle of incidence equals the angle of reflection; this is commonly stated as $>\theta_i = >\theta_r$. This behavior was first discovered through careful observation and measurement by Hero of Alexandria.> This is in contrast to diffuse reflection, where incoming light is reflected in a broad range of directions.
Visible light	The visible spectrum is the portion of the electromagnetic spectrum that is visible to the human eye. Electromagnetic radiation in this range of wavelengths is called visible light or simply light. A typical human eye will respond to wavelengths in air from about 380 to 750 nm.
Wave	A Wave is a disturbance that propagates through space and time, usually with transference of energy. A mechanical Wave is a Wave that propagates or travels through a medium due to the restoring forces it produces upon deformation. There also exist Wave s capable of traveling through a vacuum, including electromagnetic radiation and probably gravitational radiation.
Wave propagation	Wave propagation is any of the ways in which waves travel through a waveguide.

With respect to the direction of the oscillation relative to the propagation direction, we can distinguish between longitudinal wave and transverse waves.

For electromagnetic waves, propagation may occur in a vacuum as well as in a material medium.

Wavefront

In optics and physics, a wavefront is the locus of points having the same phase. Since infrared, optical, x-ray and gamma-ray frequencies are so high, the temporal component of electromagnetic waves is usually ignored at these wavelengths, and it is only the phase of the spatial oscillation that is described. Additionally, most optical systems and detectors are indifferent to polarization, so this property of the wave is also usually ignored.

Wavelength

In physics wavelength is the distance between repeating units of a propagating wave of a given frequency. It is commonly designated by the Greek letter lambda. Examples of wave-like phenomena are light, water waves, and sound waves.

White

White is not a color, the perception which is evoked by light that stimulates all three types of color sensitive cone cells in the human eye in near equal amount and with high brightness compared to the surroundings.

Since the impression of white is obtained by three summations of light intensity across the visible spectrum, the number of combinations of light wavelengths that produce the sensation of white is practically infinite. There are a number of different white light sources such as the midday Sun, incandescent lamps, fluorescent lamps and white LEDs.

White light

White is a tone, the perception which is evoked by light that stimulates all three types of color sensitive cone cells in the human eye in near equal amount and with high brightness compared to the surroundings.

Since the impression of white is obtained by three summations of light intensity across the visible spectrum, the number of combinations of light wavelengths that produce the sensation of white is practically infinite. There are a number of different white light sources such as the midday Sun, incandescent lamps, fluorescent lamps and white LEDs.

Angle

In geometry and trigonometry, an Angle (in full, plane Angle is the figure formed by two rays sharing a common endpoint, called the vertex of the Angle (Sidorov 2001.) The magnitude of the Angle is the 'amount of rotation' that separates the two rays, and can be measured by considering the length of circular arc swept out when one ray is rotated about the vertex to coincide with the other Where there is no possibility of confusion, the term Angle is used interchangeably for both the geometric configuration itself and for its angular magnitude (which is simply a numerical quantity.)

Angle of incidence

Angle of incidence is a measure of deviation of something from 'straight on', for example:

· in the approach of a ray to a surface, or
· the angle at which the wing or horizontal tail of an airplane is installed on the fuselage, measured relative to the axis of the fuselage.

In geometric optics, the Angle of incidence is the angle between a ray incident on a surface and the line perpendicular to the surface at the point of incidence, called the normal. The ray can be formed by any wave: optical, acoustic, microwave, X-ray and so on.

Euclid

Euclid , fl. 300 BC, also known as Euclid of Alexandria, was a Greek mathematician and is often referred to as the 'Father of Geometry.' He was active in Hellenistic Alexandria during the reign of Ptolemy I . His Elements is the most successful textbook and one of the most influential works in the history of mathematics, serving as the main textbook for teaching mathematics (especially geometry) from the time of its publication until the late 19th or early 20th century.

Diffraction

Diffraction is normally taken to refer to various phenomena which occur when a wave encounters an obstacle. It is described as the apparent bending of waves around small obstacles and the spreading out of waves past small openings. Similar effects are observed when light waves travel through a medium with a varying refractive index or a sound wave through one with varying acoustic impedance.

Polarization

Polarization is a property of waves that describes the orientation of their oscillations

By convention, the Polarization of light is described by specifying the direction of the wave's electric field.

Polaroid

Polaroid is the name of a type of synthetic plastic sheet which is used to polarize light.

The original material, patented in 1929 and further developed in 1932 by Edwin H. Land, consists of many microscopic crystals of iodoquinine sulfate embedded in a transparent nitrocellulose polymer film.

Liquid

Liquid is one of the principal states of matter. A Liquid is a fluid that has the particles loose and can freely form a distinct surface at the boundaries of its bulk material. The surface is a free surface where the Liquid is not constrained by a container.

Liquid crystals

Liquid crystals (Liquid crystals s) are substances that exhibit a phase of matter that has properties between those of a conventional liquid and those of a solid crystal. For instance, an Liquid crystals may flow like a liquid, but its molecules may be oriented in a crystal-like way. There are many different types of Liquid crystals phases, which can be distinguished by their different optical properties (such as birefringence.)

Parabola

In mathematics, the parabola is a conic section, the intersection of a right circular conical surface and a plane parallel to a generating straight line of that surface. Given a point and a line that lie in a plane, the locus of points in that plane that are equidistant to them is a parabola.

A particular case arises when the plane is tangent to the conical surface of a circle.

Sphere

A sphere is a symmetrical geometrical object. In non-mathematical usage, the term is used to refer either to a round ball or to its two-dimensional surface. In mathematics, a sphere is the set of all points in three-dimensional space which are at distance r from a fixed point of that space, where r is a positive real number called the radius of the sphere.

Spherical aberration

In optics, spherical aberration is a deviation from the norm resulting in an image imperfection that occurs due to the increased refraction of light rays when rays strike a lens or a reflection of light rays that occurs when rays strike a mirror near its edge, in comparison with those that strike nearer the center. A sphere lens has an aplanatic point--that means no spherical aberration--only at a radius that equals the radius of the sphere divided by the index of refraction, that is inside the lens. It is often considered to be an imperfection of telescopes and other instruments which makes their focusing less than ideal due to the spherical shape of lenses and mirrors.

Spin

In quantum mechanics, spin is a fundamental property of atomic nuclei, hadrons, and elementary particles. For particles with non-zero spin, spin direction is an important intrinsic degree of freedom.

As the name indicates, the spin has originally been thought of as a rotation of particles around their own axis.

Spin casting

Spin casting or Centrifugal Rubber Mold Casting is a method of utilizing centrifugal force to produce castings from a rubber mold. Typically, a disc-shaped mold is spun along its central axis at a set speed. The casting material, usually molten metal or liquid thermoset plastic is then poured in through an opening at the top-center of the mold.

Telescope

A telescope is an instrument designed for the observation of remote objects by the collection of electromagnetic radiation. The first known practically functioning telescope s were invented in the Netherlands at the beginning of the 17th century. ' telescope s' can refer to a whole range of instruments operating in most regions of the electromagnetic spectrum.

Large Zenith Telescope

The Large Zenith Telescope is a 6.0 m diameter liquid mirror telescope located in the University of British Columbia's Malcolm Knapp Research Forest, about 70km east of Vancouver. It is one of the largest optical telescopes in the world, but still quite inexpensive.

While a zenith telescope has the obvious disadvantage of not being able to look anywhere but a small spot straight up, its simplified setup permits the use of a mirror consisting of a smoothly spinning pan filled with liquid mercury.

Space

Space is the boundless, three-dimensional extent in which objects and events occur and have relative position and direction. Physical Space is often conceived in three linear dimensions, although modern physicists usually consider it, with time, to be part of the boundless four-dimensional continuum known as Space time. In mathematics Space s with different numbers of dimensions and with different underlying structures can be examined.

Space shuttle

NASA's Space Shuttle, officially called the Space Transportation System, is the spacecraft currently used by the United States government for its human spaceflight missions. At launch, it consists of a rust-colored external tank, two white, slender Solid Rocket Boosters, and the orbiter, a winged spaceplane which is the space shuttle in the narrow sense.

The orbiter carries astronauts and payload such as satellites or space station parts into low earth orbit, into the Earth's upper atmosphere or thermosphere.

Zenith

In broad terms, the zenith is the direction pointing directly above a particular location. Since the concept of being above is itself somewhat vague, scientists define the zenith in more rigorous terms. Specifically, in astronomy, geophysics and related sciences, the zenith at a given point is the local vertical direction pointing away from direction of the force of gravity at that location.

Adaptive optics

Adaptive optics is a technology used to improve the performance of optical systems by reducing the effects of rapidly changing optical distortion. It is used in astronomical telescopes and laser communication systems to remove the effects of atmospheric distortion, and in retinal imaging systems to reduce the impact of ocular aberrations. Adaptive optics works by measuring the distortions in a wavefront and compensating for them with a spatial phase modulator such as a deformable mirror or liquid crystal array.

W. M. Keck Observatory

The W. M. Keck Observatory is a two-telescope astronomical observatory at the 4,145 meter summit of Mauna Kea in Hawai'i. The primary mirrors of each of the two telescopes are 10 meters in diameter. The telescopes can also operate together to form a single astronomical interferometer.

Very Large Array	The Very Large Array is a radio astronomy observatory located on the Plains of San Augustin, between the towns of Magdalena and Datil, some fifty miles west of Socorro, New Mexico, USA. U.S. Route 60 passes through the complex, which is adjacent to the Boy Scout Double H High Adventure Base. The VLA stands at an altitude of 6970 ft above sea level.
Tomography	Tomography is imaging by sections or sectioning. A device used in tomography is called a tomograph, while the image produced is a tomogram. The method is used in medicine, archaeology, biology, geophysics, oceanography, materials science, astrophysics and other sciences.
Glass	Glass generally refers to hard, brittle, transparent material, such as those used for windows, many bottles, or eyewear. Examples of such solid materials include, but are not limited to, soda-lime Glass borosilicate Glass acrylic Glass sugar Glass Muscovy Glass or aluminium oxynitride. In the technical sense, Glass is an inorganic product of fusion which has been cooled through the Glass transition to a rigid condition without crystallizing.
Refraction	Refraction is the change in direction of a wave due to a change in its speed. This is most commonly seen when a wave passes from one medium to another. Refraction of light is the most commonly seen example, but any type of wave can refract when it interacts with a medium, for example when sound waves pass from one medium into another or when water waves move into water of a different depth.
Principle	A principle is one of several things: (a) a descriptive comprehensive and fundamental law, doctrine, or assumption; (b) a normative rule or code of conduct, and (c) a law or fact of nature underlying the working of an artificial device. The principle of any effect is the cause that produces it. Depending on the way the cause is understood the basic law governing that cause may acquire some distinction in its expression.
Air	The Earth's atmosphere is a layer of gases surrounding the planet Earth that is retained by Earth's gravity. The atmosphere protects life on Earth by absorbing ultraviolet solar radiation, warming the surface through heat retention (greenhouse effect), and reducing temperature extremes between day and night. Dry Air contains roughly (by volume) 78.08% nitrogen, 20.95% oxygen, 0.93% argon, 0.038% carbon dioxide, and trace amounts of other gases.
Total internal reflection	Total internal reflection is an optical phenomenon that occurs when a ray of light strikes a medium boundary at an angle larger than the critical angle with respect to the normal to the surface. If the refractive index is lower on the other side of the boundary no light can pass through, so effectively all of the light is reflected. The critical angle is the angle of incidence above which the total internal reflection occurs.
Optical fiber cable	An optical fiber cable is a cable containing one or more optical fibers. The optical fiber elements are typically individually coated with plastic layers and contained in a protective tube suitable for the environment where the cable will be deployed. In practical fibers, the cladding is usually coated with a tough resin buffer layer, which may be further surrounded by a jacket layer, usually plastic.

Optical

Optics is the science that describes the behavior and properties of light and the interaction of light with matter. Optics explains optical phenomena.

Optical fiber

An Optical fiber (or fibre) is a glass or plastic fiber that carries light along its length. Fiber optics is the overlap of applied science and engineering concerned with the design and application of Optical fiber s. Optical fiber s are widely used in fiber-optic communications, which permits transmission over longer distances and at higher bandwidths (data rates) than other forms of communications.

Optical axis

In optics, the term optical axis is used to define a direction along which there is some degree of rotational symmetry. It can be used in several contexts:

· In an optical system, the optical axis is an imaginary line that defines the path along which light propagates through the system. For a system composed of simple lenses and mirrors, the axis passes through the center of curvature of each surface, and coincides with the axis of rotational symmetry. The optical axis is often coincident with the system's mechanical axis, but not always, as in the case of off-axis optical systems.
· In an uniaxial birefringent material, the optical axis is the axis of optical anisotropy.
· For an optical fiber, the optical axis is along the center of the fiber core, and is also known as the fiber axis. .

Waves

Waves is a three-part novel by Ogan Gurel published in 2009. A 21st century version of Faust, the novel explores good and evil in both individual and global settings focusing around a hypothetical technology that has both medical and military applications. The protagonist, Tomas Twarok, is a contemplative and idealistic doctor-turned-entrepreneur who strikes a deal with his college friend, Maximilian Iblis, a ruthless hedge fund manager.

Focal length

The Focal length of an optical system is a measure of how strongly it converges (focuses) or diverges (defocuses) light. For an optical system in air, it is the distance over which initially collimated rays are brought to a focus. A system with a shorter Focal length has greater optical power than one with a long Focal length that is, it bends the rays more strongly, bringing them to a focus in a shorter distance.

Real image

In optics, a Real image is a representation of an object (source) in which the perceived location is actually a point of convergence of the rays of light that make up the image. If a screen is placed in the plane of a Real image the image will generally become visible on the screen. Examples of Real image s include the image seen on a cinema screen (the source being the projector), the image produced on a detector in the rear of a camera, and the image produced on a human retina (the latter two pass light through an internal convex lens.)

Virtual image

In optics, a virtual image is an image in which the outgoing rays from a point on the object never actually intersect at a point. A simple example is a flat mirror where the image of oneself is perceived at twice the distance from yourself to the mirror. That is, if you are half a meter in front of the mirror, your image will appear at a distance of half a meter inside or behind the mirror.

Magnification

Magnification is the process of enlarging something only in appearance, not in physical size. This enlargement is quantified by a calculated number also called Magnification When this number is less than one it refers to a reduction in size, sometimes called minification or de Magnification

Chromatic aberration

In optics, Chromatic aberration is the failure of a lens to focus all colors to the same point. It occurs because lenses have a different refractive index for different wavelengths of light The refractive index decreases with increasing wavelength.

Alhazen

AbÅ« Ê¿AlÄ« al-á¸¤asan ibn al-á¸¤asan ibn al-Haytham '>Latinized: Alhacen or Alhazen (965 in Basra - c. 1039 in Cairo) was an Arab or Persian polymath. He made significant contributions to the principles of optics, as well as to anatomy, astronomy, engineering, mathematics, medicine, ophthalmology, philosophy, physics, psychology, visual perception, and to science in general with his introduction of the scientific method.

Robert Boyle

Robert Boyle (25 January 1627 - 30 December 1691) was an Irish theologian, natural philosopher, chemist, physicist, inventor, and early gentleman scientist, noted for his work in physics and chemistry. He is best known for the formulation of Boyle's law. Although his research and personal philosophy clearly has its roots in the alchemical tradition, he is largely regarded today as the first modern chemist, and therefore one of the founders of modern chemistry.

Orbit

In physics, an orbit is the gravitationally curved path of one object around a point or another body, for example the gravitational orbit of a planet around a star.

Historically, the apparent motion of the planets were first understood in terms of epicycles, which are the sums of numerous circular motions. This predicted the path of the planets quite well, until Johannes Kepler was able to show that the motion of the planets were in fact elliptical motions.

Orbits

The velocity relationship of two objects with mass can thus be considered in four practical classes, with subtypes:

· No orbit
· Interrupted orbits

· Range of interrupted elliptical paths
· Circumnavigating orbits

· Range of elliptical paths with closest point opposite firing point
· Circular path
· Range of elliptical paths with closest point at firing point
· Infinite orbits

· Parabolic paths
· Hyperbolic paths

In many situations relativistic effects can be neglected, and Newton's laws give a highly accurate description of the motion. Then the acceleration of each body is equal to the sum of the gravitational forces on it, divided by its mass, and the gravitational force between each pair of bodies is proportional to the product of their masses and decreases inversely with the square of the distance between them. To this Newtonian approximation, for a system of two point masses or spherical bodies, only influenced by their mutual gravitation, the orbits can be exactly calculated. If the heavier body is much more massive than the smaller, as for a satellite or small moon orbiting a planet or for the Earth orbiting the Sun, it is accurate and convenient to describe the motion in a coordinate system that is centered on the heavier body, and we can say that the lighter body is in orbit around the heavier.

Satellite

In the context of spaceflight, a satellite is an object which has been placed into orbit by human endeavor. Such objects are sometimes called artificial satellite s to distinguish them from natural satellite s such as the Moon.

The first artificial satellite Sputnik 1, was launched by the Soviet Union in 1957.

Eye

Eye s are organs that detect light, and send electrical impulses along the optic nerve to the visual and other areas of the brain. Complex optical systems with resolving power have come in ten fundamentally different forms, and 96% of animal species possess a complex optical system. Image-resolving Eye s are present in cnidaria, molluscs, chordates, annelids and arthropods.

Photorefractive keratectomy

Photorefractive keratectomy and Laser-Assisted Sub-Epithelial Keratectomy are laser eye surgery procedures intended to correct a person's vision, reducing dependency on glasses or contact lenses. The first LASEK procedure was performed at Massachusetts Eye and Ear Infirmary in 1996 by ophthalmologist, refractive surgeon, Dimitri Azar. The procedure was later popularized by Camellin, who coined the term LASEK for laser epithelial keratomileusis.

Radial keratotomy

Radial keratotomy is a refractive surgical procedure to correct myopia.

The procedure was discovered by accident by Svyatoslav Fyodorov who removed glass from the eye of one of his patients who had been in an accident. A boy, who wore eyeglasses, fell off his bicycle and his glasses shattered on impact, with glass particles lodging in his eyes.

Radiation

Radiation, as in physics, is energy in the form of waves or moving subatomic particles emitted by an atom or other body as it changes from a higher energy state to a lower energy state. Radiation can be classified as ionizing or non-ionizing radiation, depending on its effect on atomic matter. The most common use of the word 'radiation' refers to ionizing radiation.

Rainbow

A rainbow is an optical and meteorological phenomenon that causes a spectrum of light to appear in the sky when the Sun shines onto droplets of moisture in the Earth's atmosphere. They take the form of a multicoloured arc, with red on the outer part of the arch and violet on the inner section of the arch.

More rarely, a secondary rainbow is seen, which is a second, fainter arc, outside the primary arc, with colours in the opposite order, that is, with violet on the outside and red on the inside.

Angular momentum

Angular momentum is a quantity that is useful in describing the rotational state of a physical system. For a rigid body rotating around an axis of symmetry (e.g. the fins of a ceiling fan), the Angular momentum can be expressed as the product of the body's moment of inertia and its angular velocity ($\mathbf{L} = I\boldsymbol{\omega}$.) In this way, Angular momentum is sometimes described as the rotational analog of linear momentum.

Crystal

A Crystal or Crystal line solid is a solid material whose constituent atoms, molecules, or ions are arranged in an orderly repeating pattern extending in all three spatial dimensions. The scientific study of Crystal s and Crystal formation is Crystal lography. The process of Crystal formation via mechanisms of Crystal growth is called Crystal lization or solidification.

Linear

The word Linear comes from the Latin word Linear is, which means created by lines. In mathematics, a Linear map or function f(x) is a function which satisfies the following two properties...

· Additivity (also called the superposition property): f(x + y) = f(x) + f(y.)

Corona

A Corona is a type of plasma 'atmosphere' of the Sun or other celestial body, extending millions of kilometers into space, most easily seen during a total solar eclipse, but also observable in a Corona graph. The Latin root of the word Corona means crown. The Corona can still be seen in this solar eclipse of August 11, 1999, as seen from France.

The high temperature of the Corona gives it unusual spectral features, which led some to suggest, in the 19th century, that it contained a previously unknown element, 'coronium'.

Electric charge

Electric charge is a fundamental conserved property of some subatomic particles, which determines their electromagnetic interaction. Electrically charged matter is influenced by, and produces, electromagnetic fields. The interaction between a moving charge and an electromagnetic field is the source of the electromagnetic force, which is one of the four fundamental forces.

Electric field

In physics, the space surrounding an electric charge or in the presence of a time-varying magnetic field has a property called an Electric field This Electric field exerts a force on other electrically charged objects. The concept of an Electric field was introduced by Michael Faraday.

Radiant

The radiant of a meteor shower is the point in the sky that meteors appear to originate from. For example, the Perseids are meteors which appear to come from a point within the constellation of Perseun.

Radiant energy

Radiant energy is the energy of electromagnetic waves. The quantity of Radiant energy may be calculated by integrating radiant flux (or power) with respect to time and, like all forms of energy, its SI unit is the joule. The term is used particularly when radiation is emitted by a source into the surrounding environment.

Ozone

Ozone or trioxygen is a triatomic molecule, consisting of three oxygen atoms. It is an allotrope of oxygen that is much less stable than the diatomic O_2. Ground-level ozone is an air pollutant with harmful effects on the respiratory systems of animals and humans.

Ozone layer

The photochemical mechanisms that give rise to the ozone layer were worked out by the British physicist Sidney Chapman in 1930. Ozone in the earth's stratosphere is created by ultraviolet light striking oxygen molecules containing two oxygen atoms, splitting them into individual oxygen atoms; surface. About 90% of the ozone in our atmosphere is contained in the stratosphere.

Venus

Venus is the second-closest planet to the Sun, orbiting it every 224.7 Earth days. The planet is named after Venus, the Roman goddess of love. It is the brightest natural object in the night sky, except for the Moon, reaching an apparent magnitude of -4.6.

Refracting

A refracting or refractor telescope is a dioptric telescope that uses a lens as its objective to form an image. The refracting telescope design was originally used in spy glasses and astronomical telescopes but is also used in other devices such as binoculars and long or telephoto camera lenses.

Refractors were the earliest type of optical telescope.

Refracting Telescope

A refracting is a dioptric telescope that uses a lens as its objective to form an image. The refracting telescope design was originally used in spy glasses and astronomical telescopes but is also used in other devices such as binoculars and long or telephoto camera lenses.

Refractors were the earliest type of optical telescope.

Reflecting Telescope

A Reflecting telescope is an optical telescope which uses a single or combination of curved mirrors that reflect light and form an image. The Reflecting telescope was invented in the 17th century as an alternative to the refracting telescope which, at that time, was a design that suffered from severe chromatic aberration. Although Reflecting telescope s produce other types of optical aberrations, it is a design that allows for very large diameter objectives.

Electromagnetism

Electromagnetism is the physics of the electromagnetic field, a field that exerts a force on particles with the property of electric charge and is reciprocally affected by the presence and motion of such particles.

A changing magnetic field produces an electric field (this is the phenomenon of electromagnetic induction, the basis of operation for electrical generators, induction motors, and transformers.) Similarly, a changing electric field generates a magnetic field.

Gauss

Johann Carl Friedrich Gauss , Latin: Carolus Fridericus Gauss /span>) was a German mathematician and scientist who contributed significantly to many fields, including number theory, statistics, analysis, differential geometry, geodesy, geophysics, electrostatics, astronomy and optics. Sometimes known as the Princeps mathematicorum and 'greatest mathematician since antiquity', Gauss had a remarkable influence in many fields of mathematics and science and is ranked as one of history's most influential mathematicians. He referred to mathematics as 'the queen of sciences.'

Magnetism

In physics, the term Magnetism is used to describe how materials respond on the microscopic level to an applied magnetic field; to categorize the magnetic phase of a material. For example, the most well known form of Magnetism is ferro Magnetism such that some ferromagnetic materials produce their own persistent magnetic field. Some well-known ferromagnetic materials that exhibit easily detectable magnetic properties (to form magnets) are nickel, iron, cobalt, gadolinium and their alloys.

Atomic physics	Atomic physics (or atom physics) is the field of physics that studies atoms as an isolated system of electrons and an atomic nucleus. It is primarily concerned with the arrangement of electrons around the nucleus and the processes by which these arrangements change. This includes ions as well as neutral atoms and, unless otherwise stated, for the purposes of this discussion it should be assumed that the term atom includes ions.
Plasma	In physics and chemistry, plasma is an ionized gas, in which a certain proportion of electrons are free rather than being bound to an atom or molecule. The ability of the positive and negative charges to move somewhat independently makes the plasma electrically conductive so that it responds strongly to electromagnetic fields. Plasma therefore has properties quite unlike those of solids, liquids or gases and is considered to be a distinct state of matter.
Quantum	In physics, a quantum is an indivisible entity of a quantity that has the same units as the Planck constant and is related to both energy and momentum of elementary particles of matter and of photons and other bosons. The word comes from the Latin 'quantus,' for 'how much.' Behind this, one finds the fundamental notion that a physical property may be 'quantized', referred to as 'quantization'. This means that the magnitude can take on only certain discrete numerical values, rather than any value, at least within a range.
Photoelectric effect	The photoelectric effect is a quantum electronic phenomenon in which electrons are emitted from matter after the absorption of energy from electromagnetic radiation such as x-rays or visible light. The emitted electrons can be referred to as photoelectrons in this context. The effect is also termed the Hertz Effect, due to its discovery by Heinrich Rudolf Hertz, although the term has generally fallen out of use.
X-radiation	X-radiation is a form of electromagnetic radiation. X-rays have a wavelength in the range of 10 to 0.01 nanometers, corresponding to frequencies in the range 30 petahertz to 30 exahertz and energies in the range 120 eV to 120 keV. They are longer than gamma rays but shorter than UV rays.
Radiation	Radiation, as in physics, is energy in the form of waves or moving subatomic particles emitted by an atom or other body as it changes from a higher energy state to a lower energy state. Radiation can be classified as ionizing or non-ionizing radiation, depending on its effect on atomic matter. The most common use of the word 'radiation' refers to ionizing radiation.
Wave	A Wave is a disturbance that propagates through space and time, usually with transference of energy. A mechanical Wave is a Wave that propagates or travels through a medium due to the restoring forces it produces upon deformation. There also exist Wave s capable of traveling through a vacuum, including electromagnetic radiation and probably gravitational radiation.
Waves	Waves is a three-part novel by Ogan Gurel published in 2009. A 21st century version of Faust, the novel explores good and evil in both individual and global settings focusing around a hypothetical technology that has both medical and military applications. The protagonist, Tomas Twarok, is a contemplative and idealistic doctor-turned-entrepreneur who strikes a deal with his college friend, Maximilian Iblis, a ruthless hedge fund manager.
Wavelength	In physics wavelength is the distance between repeating units of a propagating wave of a given frequency. It is commonly designated by the Greek letter lambda. Examples of wave-like phenomena are light, water waves, and sound waves.

Energy

In physics, Energy is a scalar physical quantity that describes the amount of work that can be performed by a force, an attribute of objects and systems that is subject to a conservation law. Different forms of Energy include kinetic, potential, thermal, gravitational, sound, light, elastic, and electromagnetic Energy The forms of Energy are often named after a related force.

Orbit

In physics, an orbit is the gravitationally curved path of one object around a point or another body, for example the gravitational orbit of a planet around a star.

Historically, the apparent motion of the planets were first understood in terms of epicycles, which are the sums of numerous circular motions. This predicted the path of the planets quite well, until Johannes Kepler was able to show that the motion of the planets were in fact elliptical motions.

Orbits

The velocity relationship of two objects with mass can thus be considered in four practical classes, with subtypes:

- No orbit
- Interrupted orbits

- Range of interrupted elliptical paths
- Circumnavigating orbits

- Range of elliptical paths with closest point opposite firing point
- Circular path
- Range of elliptical paths with closest point at firing point
- Infinite orbits

- Parabolic paths
- Hyperbolic paths

In many situations relativistic effects can be neglected, and Newton's laws give a highly accurate description of the motion. Then the acceleration of each body is equal to the sum of the gravitational forces on it, divided by its mass, and the gravitational force between each pair of bodies is proportional to the product of their masses and decreases inversely with the square of the distance between them. To this Newtonian approximation, for a system of two point masses or spherical bodies, only influenced by their mutual gravitation, the orbits can be exactly calculated. If the heavier body is much more massive than the smaller, as for a satellite or small moon orbiting a planet or for the Earth orbiting the Sun, it is accurate and convenient to describe the motion in a coordinate system that is centered on the heavier body, and we can say that the lighter body is in orbit around the heavier.

Planck

Max Planck (April 23, 1858 - October 4, 1947) was a German physicist. He is considered to be the founder of the quantum theory, and thus one of the most important physicists of the twentieth century. Planck was awarded the Nobel Prize in Physics in 1918.

Potential energy

Potential energy can be thought of as energy stored within a physical system. It is called potential energy because it has the potential to be converted into other forms of energy, such as kinetic energy, and to do work in the process. The standard unit of measure for potential energy is the joule, the same as for work, or energy in general.

Quantization

In physics, Quantization is a procedure for constructing a quantum field theory starting from a classical field theory. This is a generalization of the procedure for building quantum mechanics from classical mechanics. One also speaks of field Quantization, as in the 'Quantization of the electromagnetic field', where one refers to photons as field 'quanta' (for instance as light quanta.)

Quantum mechanics

Quantum mechanics is the study of mechanical systems whose dimensions are close to the atomic scale, such as molecules, atoms, electrons, protons and other subatomic particles. Quantum mechanics is a fundamental branch of physics with wide applications. Quantum theory generalizes classical mechanics to provide accurate descriptions for many previously unexplained phenomena such as black body radiation and stable electron orbits.

Big Bang

The Big Bang is the cosmological model of the initial conditions and subsequent development of the universe that is supported by the most comprehensive and accurate explanations from current scientific evidence and observation. As used by cosmologists, the term Big Bang generally refers to the idea that the universe has expanded from a primordial hot and dense initial condition at some finite time in the past (currently estimated to have been approximately 13.7 billion years ago), and continues to expand to this day.

Georges Lemaître proposed what became known as the Big Bang theory of the origin of the Universe, although he called it his 'hypothesis of the primeval atom'.

Elementary particle

In particle physics, an Elementary particle or fundamental particle is a particle not known to have substructure; that is, it is not known to be made up of smaller particles. If an Elementary particle truly has no substructure, then it is one of the basic building blocks of the universe from which all other particles are made. In the Standard Model, the quarks, leptons, and gauge bosons are Elementary particle s.

Photon

In physics, the photon is the elementary particle responsible for electromagnetic phenomena. It is the carrier of electromagnetic radiation of all wavelengths, including in decreasing order of energy, gamma rays, X-rays, ultraviolet light, visible light, infrared light, microwaves, and radio waves. The photon differs from many other elementary particles, such as the electron and the quark, in that it has zero rest mass; therefore, it travels at the speed of light, c.

Spark

In mathematics, specifically in linear algebra, the Spark of a matrix A is the smallest number n such that there exists a set of n columns in A which are linearly dependent. Formally,

$$\text{spark}(A) = \min_{d \neq 0} \|d\|_0 \text{ s.t. } Ad = 0.$$

By contrast, the rank of a matrix is the smallest number k such that all sets of k + 1 columns in A are linearly dependent. The concept of the Spark is of use in the theory of compressive sensing, where requirements on the Spark of the measurement matrix are used to ensure stability and consistency of various estimation techniques.

Ultraviolet

Ultraviolet light is electromagnetic radiation with a wavelength shorter than that of visible light, but longer than X-rays. It is so named because the spectrum consists of electromagnetic waves with frequencies higher than those that humans identify as the color violet.

UV light is typically found as part of the radiation received by the Earth from the Sun.

Visible light

The visible spectrum is the portion of the electromagnetic spectrum that is visible to the human eye. Electromagnetic radiation in this range of wavelengths is called visible light or simply light. A typical human eye will respond to wavelengths in air from about 380 to 750 nm.

Voltage

Voltage is commonly used as a short name for electrical potential difference. In this introduction, the term 'Voltage' is however used to mean electric potential, i.e. a hypothetically measurable physical dimension, and is denoted by the algebraic variable V (inclined or italicized letter.) The SI unit for Voltage is the volt (symbol: V [not italicized].)

Gamma rays

Gamma rays (denoted as γ) are electromagnetic radiation of high energy. They are produced by sub-atomic particle interactions, such as electron-positron annihilation, neutral pion decay, radioactive decay, fusion, fission or inverse Compton scattering in astrophysical processes. Gamma rays typically have frequencies above 10^{19} Hz and therefore energies above 100 keV and wavelength less than 10 picometers, often smaller than an atom.

Radio

Radio is the transmission of signals, by modulation of electromagnetic waves with frequencies below those of visible light. Electromagnetic radiation travels by means of oscillating electromagnetic fields that pass through the air and the vacuum of space. Information is carried by systematically changing some property of the radiated waves, such as amplitude, frequency, or phase.

Radio waves

Radio waves are electromagnetic waves occurring on the radio frequency portion of the electromagnetic spectrum. A common use is to transport information through the atmosphere or outer space without wires. Radio waves are distinguished from other kinds of electromagnetic waves by their wavelength, a relatively long wavelength in the electromagnetic spectrum.

Television

Television is a widely used telecommunication medium for sending and receiving moving images, either monochromatic or color, usually accompanied by sound. 'Television' may also refer specifically to a television set, television programming or television transmission. The word is derived from mixed Latin and Greek roots, meaning 'far sight': Greek tele, far, and Latin visio, sight.

Atmosphere

An Atmosphere is a layer of gases that may surround a material body of sufficient mass, by the gravity of the body, and are retained for a longer duration if gravity is high and the Atmosphere s temperature is low. Some planets consist mainly of various gases, but only their outer layer is their Atmosphere .

The term stellar Atmosphere describes the outer region of a star, and typically includes the portion starting from the opaque photosphere outwards.

Spectrum

A spectrum is a condition that is not limited to a specific set of values but can vary infinitely within a continuum. The word saw its first scientific use within the field of optics to describe the rainbow of colors in visible light when separated using a prism; it has since been applied by analogy to many fields other than optics. Thus, one might talk about the spectrum of political opinion, or the spectrum of activity of a drug, or the autism spectrum.

Astronomy	Astronomy , 'law') is the scientific study of celestial objects (such as stars, planets, comets, and galaxies) and phenomena that originate outside the Earth's atmosphere (such as the cosmic background radiation.) It is concerned with the evolution, physics, chemistry, meteorology, and motion of celestial objects, as well as the formation and development of the universe.
Electricity	Electricity is a general term that encompasses a variety of phenomena resulting from the presence and flow of electric charge. These include many easily recognizable phenomena, such as lightning and static Electricity but in addition, less familiar concepts, such as the electromagnetic field and electromagnetic induction. In general usage, the word Electricity is adequate to refer to a number of physical effects.
Xerography	Xerography is a photocopying technique developed by Chester Carlson in 1938 and patented on October 6, 1942. He received U.S. Patent 2,297,691 for his invention. Although dry electrostatic printing processes had been invented as far back as 1778 by Georg Christoph Lichtenberg, Carlson's innovation combined electrostatic printing with photography.
Electrostatic	Electrostatic s is the branch of science that deals with the phenomena arising from stationary or slowly moving electric charges. Since classical antiquity it was known that some materials such as amber attract light particles after rubbing. The Greek word for amber, Î®λεκτρον , was the source of the word 'electricity'.
Photoconductivity	Photoconductivity is an optical and electrical phenomenon in which a material becomes more conductive due to the absorption of electro-magnetic radiation such as visible light, ultraviolet light, infrared light, or gamma radiation. Lead content of the surrounding area can be a factor in the effectiveness of this, however. When light is absorbed by a material like a semiconductor, the number of free electrons and holes changes and raises the electrical conductivity of the semiconductor.
Semiconductor	A semiconductor is a material that has an electrical resistivity between that of a conductor and an insulator. An external electrical field changes a semiconductor s resistivity. Devices made from semiconductor materials are the foundation of modern electronics, including radio, computers, telephones, and many other devices.
Solar	Solar power uses Solar Radiation emitted from our sun. Solar power, a renewable energy source, has been used in many traditional technologies for centuries, and is in widespread use where other power supplies are absent, such as in remote locations and in space.
Solar energy	Solar energy is the light and radiant heat from the Sun that influences Earth's climate and weather and sustains life. Solar power is the rate of solar energy at a point in time; it is sometimes used as a synonym for solar energy or more specifically to refer to electricity generated from solar radiation. Since ancient times solar energy has been harnessed for human use through a range of technologies.
Star	A star is a massive, luminous ball of plasma that is held together by gravity. The nearest star to Earth is the Sun, which is the source of most of the energy on Earth. Other star s are visible in the night sky, when they are not outshone by the Sun.

Transmitter	A transmitter is an electronic device which, usually with the aid of an antenna, propagates an electromagnetic signal such as radio, television, or other telecommunications. In other applications signals can also be transmitted using an analog 0/4-20 mA current loop signal. WDET-FM transmitter Generally and in communication and information processing, a transmitter is any object which sends information to an observer.
Gas	In physics, a Gas is a state of matter, consisting of a collection of particles (molecules, atoms, ions, electrons, etc.) without a definite shape or volume that are in more or less random motion. Gas ses are generally modeled by the ideal Gas law where the total volume of the Gas increases proportionally to absolute temperature and decreases inversely proportionally to pressure.
Periodic table	The periodic table of the chemical elements is a tabular method of displaying the chemical elements. Although precursors to this table exist, its invention is generally credited to Russian chemist Dmitri Mendeleev in 1869. Mendeleev intended the table to illustrate recurring trends in the properties of the elements.
Solid	The solid state of matter is characterized by a distinct structural rigidity and virtual resistance to deformation (i.e. changes of shape and/or volume.) Most solid s have high values both of Young's modulus and of the shear modulus of elasticity. This contrasts with most liquids or fluids, which have a low shear modulus, and typically exhibit the capacity for macroscopic viscous flow.
Robert Wilhelm Bunsen	Robert Wilhelm Bunsen was a German chemist. His laboratory assistant, Peter Desaga perfected the burner that was later named after Bunsen, which was invented originally by British chemist/physicist Michael Faraday. He also worked on emission spectroscopy of heated elements. Together, he and Gustav Kirchhoff discovered the elements cesium and rubidium. He is considered to be the developer of modern gas-analytical methods.
Spectroscopy	Spectroscopy was originally the study of the interaction between radiation and matter as a function of wavelength. In fact, historically, spectroscopy referred to the use of visible light dispersed according to its wavelength, e.g. by a prism. Later the concept was expanded greatly to comprise any measurement of a quantity as function of either wavelength or frequency.
Telescope	A telescope is an instrument designed for the observation of remote objects by the collection of electromagnetic radiation. The first known practically functioning telescope s were invented in the Netherlands at the beginning of the 17th century. ' telescope s' can refer to a whole range of instruments operating in most regions of the electromagnetic spectrum.
Solar system	The Solar System consists of the Sun and those celestial objects bound to it by gravity. These objects are the eight planets, their 166 known moons, five dwarf planets, and billions of small bodies. The small bodies include asteroids, icy Kuiper belt objects, comets, meteoroids, and interplanetary dust.
Electron	An Electron is a subatomic particle that carries a negative electric charge. It has no known substructure and is believed to be a point particle. An Electron has a mass that is approximately 1836 times less than that of the proton.
Ionization	Ionization is the physical process of converting an atom or molecule into an ion by adding or removing charged particles such as electrons or other ions. This is often confused with dissociation (chemistry.)

The process works slightly differently depending on whether an ion with a positive or a negative electric charge is being produced.

Absorption spectrum	A material's Absorption spectrum shows the fraction of incident electromagnetic radiation absorbed by the material over a range of frequencies. An Absorption spectrum is, in a sense, the opposite of an emission spectrum. Every chemical element has absorption lines at several particular wavelengths corresponding to the differences between the energy levels of its atomic orbitals.
Acceleration	In physics, and more specifically kinematics, Acceleration is the change in velocity over time. Because velocity is a vector, it can change in two ways: a change in magnitude and/or a change in direction. In one dimension, Acceleration is the rate at which something speeds up or slows down.
Angular momentum	Angular momentum is a quantity that is useful in describing the rotational state of a physical system. For a rigid body rotating around an axis of symmetry (e.g. the fins of a ceiling fan), the Angular momentum can be expressed as the product of the body's moment of inertia and its angular velocity ($\mathbf{L} = I\boldsymbol{\omega}$.) In this way, Angular momentum is sometimes described as the rotational analog of linear momentum.
Diffraction	Diffraction is normally taken to refer to various phenomena which occur when a wave encounters an obstacle. It is described as the apparent bending of waves around small obstacles and the spreading out of waves past small openings. Similar effects are observed when light waves travel through a medium with a varying refractive index or a sound wave through one with varying acoustic impedance.
Magnification	Magnification is the process of enlarging something only in appearance, not in physical size. This enlargement is quantified by a calculated number also called Magnification When this number is less than one it refers to a reduction in size, sometimes called minification or de Magnification
Molecule	A Molecule is defined as a sufficiently stable, electrically neutral group of at least two atoms in a definite arrangement held together by very strong (covalent) chemical bonds. Molecule s are distinguished from polyatomic ions in this strict sense. In organic chemistry and biochemistry, the term Molecule is used less strictly and also is applied to charged organic Molecule s and bio Molecule s.
Scanning tunneling microscope	Scanning tunneling microscope is a powerful technique for viewing surfaces at the atomic level. Its development in 1981 earned its inventors, Gerd Binnig and Heinrich Rohrer, the Nobel Prize in Physics in 1986 . STM probes the density of states of a material using tunneling current.
Werner Heisenberg	Werner Heisenberg was a German theoretical physicist, best known for enunciating the uncertainty principle of quantum theory. He made important contributions to quantum mechanics, nuclear physics, quantum field theory, and particle physics. Heisenberg, along with Max Born and Pascual Jordan, set forth the matrix formulation of quantum mechanics in 1925.
Principle	A principle is one of several things: (a) a descriptive comprehensive and fundamental law, doctrine, or assumption; (b) a normative rule or code of conduct, and (c) a law or fact of nature underlying the working of an artificial device. The principle of any effect is the cause that produces it.

Depending on the way the cause is understood the basic law governing that cause may acquire some distinction in its expression.

Energy levels

A quantum mechanical system or particle that is bound, confined spatially, can only take on certain discrete values of energy, as opposed to classical particles, which can have any energy. These values are called Energy levels The term is most commonly used for the Energy levels of electrons in atoms or molecules, which are bound by the electric field of the nucleus.

Wave function

A wave function or wavefunction is a mathematical tool used in quantum mechanics to describe any physical system. It is a function from a space that maps the possible states of the system into the complex numbers. The laws of quantum mechanics describe how the wave function evolves over time.

Stationary state

In quantum mechanics, a stationary state is an eigenstate of a Hamiltonian, or in other words, a state of definite energy. It is called stationary because the corresponding probability density has no time dependence.

As an eigenstate of the Hamiltonian, a stationary state is not subject to change or decay.

Wolfgang Ernst Pauli

Wolfgang Ernst Pauli (April 25, 1900 - December 15, 1958) was an Austrian theoretical physicist noted for his work on spin theory, and for the discovery of the exclusion principle underpinning the structure of matter and the whole of chemistry.

Quantum numbers

Quantum numbers describe values of conserved numbers in the dynamics of the quantum system. They often describe specifically the energies of electrons in atoms, but other possibilities include angular momentum, spin etc. Since any quantum system can have one or more quantum numbers, it is a futile job to list all possible quantum numbers.

Balmer series

The Balmer series or Balmer lines in atomic physics, is the designation of one of a set of six different named series describing the spectral line emissions of the hydrogen atom. The Balmer series is calculated using the Balmer formula, an empirical equation discovered by Johann Balmer in 1885.

The visible spectrum of light from hydrogen displays four wavelengths, 410 nm, 434 nm, 486 nm, and 656 nm, that reflect emissions of photons by electrons in excited states transitioning to the quantum level described by the principal quantum number n equals 2.

Collision

A Collision is an isolated event in which two or more moving bodies (colliding bodies) exert relatively strong forces on each other for a relatively short time. Deflection happens when an object hits a plane surface

Collision s involve forces (there is a change in velocity.) Collision s can be elastic, meaning they conserve energy and momentum, inelastic, meaning they conserve momentum but not energy, or totally inelastic (or plastic), meaning they conserve momentum and the two objects stick together.

Paschen series

In physics, the Paschen series is the series of transitions and resulting emission lines of the hydrogen atom as an electron goes from $n \geq 4$ to $n = 3$, where n refers to the principal quantum number of the electron. The transitions are named sequentially by Greek letter: $n = 4$ to $n = 3$ is called Paschen-alpha, 5 to 3 is Paschen-beta, 6 to 3 is Paschen-gamma, etc.

They are named after the Austro-German physicist Friedrich Paschen who first observed them in 1908.

Atom

The Atom is a basic unit of matter consisting of a dense, central nucleus surrounded by a cloud of negatively charged electrons. The Atom ic nucleus contains a mix of positively charged protons and electrically neutral neutrons (except in the case of hydrogen-1, which is the only stable nuclide with no neutron.) The electrons of an Atom are bound to the nucleus by the electromagnetic force.

Kinetic energy

The Kinetic energy of an object is the extra energy which it possesses due to its motion. It is defined as the work needed to accelerate a body of a given mass from rest to its current velocity. Having gained this energy during its acceleration, the body maintains this Kinetic energy unless its speed changes.

Photoionisation

Photoionisation is the physical process in which an incident photon ejects one or more electrons from an atom, ion or molecule.

The ejected electrons, known as photoelectrons, carry information about their pre-ionised states. For example, a single electron can have a kinetic energy equal to the energy of the incident photon minus the electron binding energy of the state it left.

White

White is not a color, the perception which is evoked by light that stimulates all three types of color sensitive cone cells in the human eye in near equal amount and with high brightness compared to the surroundings.

Since the impression of white is obtained by three summations of light intensity across the visible spectrum, the number of combinations of light wavelengths that produce the sensation of white is practically infinite. There are a number of different white light sources such as the midday Sun, incandescent lamps, fluorescent lamps and white LEDs.

White light

White is a tone, the perception which is evoked by light that stimulates all three types of color sensitive cone cells in the human eye in near equal amount and with high brightness compared to the surroundings.

Since the impression of white is obtained by three summations of light intensity across the visible spectrum, the number of combinations of light wavelengths that produce the sensation of white is practically infinite. There are a number of different white light sources such as the midday Sun, incandescent lamps, fluorescent lamps and white LEDs.

Frequency

Frequency is the number of occurrences of a repeating event per unit time. It is also referred to as temporal Frequency The period is the duration of one cycle in a repeating event, so the period is the reciprocal of the Frequency

Magnetic field

Magnetic field s surround magnetic materials and electric currents and are detected by the force they exert on other magnetic materials and moving electric charges. The Magnetic field at any given point is specified by both a direction and a magnitude (or strength); as such it is a vector field.

For the physics of magnetic materials, see magnetism and magnet, more specifically ferromagnetism, paramagnetism, and diamagnetism.

Nitrogen

Nitrogen is a chemical element that has the symbol N and atomic number 7 and atomic weight 14.0067. Elemental nitrogen is a colorless, odorless, tasteless and mostly inert diatomic gas at standard conditions, constituting 80% by volume of Earth's atmosphere.

Many industrially important compounds, such as ammonia, nitric acid, organic nitrates, and cyanides, contain nitrogen.

North pole	The North Pole is, subject to the caveats explained below, defined as the point in the northern hemisphere where the Earth's axis of rotation meets the Earth's surface. It should not be confused with the North Magnetic Pole. The North Pole is the northernmost point on Earth, lying diametrically opposite the South Pole.
Oxygen	Oxygen and -γενÎ®ς (-genÄ“s) (producer, literally begetter) is the element with atomic number 8 and represented by the symbol O. It is a member of the chalcogen group on the periodic table, and is a highly reactive nonmetallic period 2 element that readily forms compounds (notably oxides) with almost all other elements. At standard temperature and pressure two atoms of the element bind to form di Oxygen , a colorless, odorless, tasteless diatomic gas with the formula O_2. Oxygen is the third most abundant element in the universe by mass after hydrogen and helium and the most abundant element by mass in the Earth's crust.
Phosphor	A Phosphor is a substance that exhibits the phenomenon of Phosphor escence (sustained glowing after exposure to energized particles such as electrons or ultraviolet photons.) Phosphor s are transition metal compounds or rare earth compounds of various types. The most common uses of Phosphor s are in CRT displays and fluorescent lights.
Solar wind	The solar wind is a stream of charged particles--a plasma--that are ejected from the upper atmosphere of the sun. It consists mostly of electrons and protons with energies of about 1 keV. These particles are able to escape the sun's gravity, in part because of the high temperature of the corona, but also because of high kinetic energy that particles gain through a process that is not well-understood at this time.
South pole	The South Pole is the southernmost point on the surface of the Earth. It lies on the continent of Antarctica, on the opposite side of the Earth from the North Pole. It is the site of the United States Amundsen-Scott South Pole Station, which was established in 1956 and has been permanently staffed since that year.
Sodium	Sodium is an element which has the symbol N, atomic number 11, atomic mass 22.9898 g/mol, common oxidation number +1. Sodium is a soft, silvery white, highly reactive element and is a member of the alkali metals within 'group 1'. It has only one stable isotope, ^{23}Na.
Bremsstrahlung	Bremsstrahlung , is electromagnetic radiation produced by the acceleration of a charged particle, such as an electron, when deflected by another charged particle, such as an atomic nucleus. The term is also used to refer to the process of producing the radiation. Bremsstrahlung has a continuous spectrum.
Tungsten	Tungsten is a chemical element that has the symbol W and atomic number 74. A steel-gray metal, tungsten is found in several ores, including wolframite and scheelite. It is remarkable for its robust physical properties, especially the fact that it has the highest melting point of all the non-alloyed metals and the second highest of all the elements after carbon.
X-rays	X-radiation is a form of electromagnetic radiation. X-rays have a wavelength in the range of 10 to 0.01 nanometers, corresponding to frequencies in the range 30 petahertz to 30 exahertz and energies in the range 120 eV to 120 keV. They are longer than gamma rays but shorter than UV rays.

K-alpha	In X-ray spectroscopy, K-alpha emission lines result when an electron transitions to the innermost 'K' shell (principal quantum number 1) from a 2p orbital of the second or 'L' shell (with principal quantum number 2.) The line is actually a doublet, with slightly different energies depending on spin-orbit interaction energy between the electron spin and the orbital momentum of the 2p orbital. K-alpha is typically by far the strongest X-ray spectral line for an element bombarded with energy sufficient to cause maximally intense X-ray emission.
K-beta	K-beta emissions, similar to K-alpha emissions, result when an electron transitions to the innermost 'K' shell (principal quantum number 1) from a 3p orbital of the second or 'M' shell (with principal quantum number 2.) Values can be found in the X-Ray Transition Energies Database.
Population	In biology a population is the collection of inter-breeding organisms of a particular species; in sociology, a collection of human beings. A population shares a particular characteristic of interest, most often that of living in a given geographic area. In taxonomy population is a low-level taxonomic rank.
Population inversion	In physics, specifically statistical mechanics, a population inversion occurs when a system exists in state with more members in an excited state than in lower energy states. The concept is of fundamental importance in laser science because the production of a population inversion is a necessary step in the workings of a laser. To understand the concept of a population inversion, it is necessary to understand some thermodynamics and the way that light interacts with matter.
Mirror	A Mirror or looking glass is an object with at least one polished and therefore specularly reflective surface. The most familiar type of Mirror is the plane Mirror which has a flat surface. Curved Mirror s are also used, to produce magnified or diminished images or focus light or simply distort the reflected image.
Optical	Optics is the science that describes the behavior and properties of light and the interaction of light with matter. Optics explains optical phenomena.
Optical pumping	Optical pumping is a process in which light is used to raise electrons from a lower energy level in an atom or molecule to a higher one. It is commonly used in laser construction, to pump the active laser medium so as to achieve population inversion. The technique was developed by 1966 Nobel Prize winner Alfred Kastler in the early 1950s.
Optics	Optics is the science that describes the behavior and properties of light and the interaction of light with matter. Optics explains optical phenomena. The word optics comes from á½€πτικÎ®, meaning appearance or look in Ancient Greek.)
Crystal	A Crystal or Crystal line solid is a solid material whose constituent atoms, molecules, or ions are arranged in an orderly repeating pattern extending in all three spatial dimensions. The scientific study of Crystal s and Crystal formation is Crystal lography. The process of Crystal formation via mechanisms of Crystal growth is called Crystal lization or solidification.
Optical fiber cable	An optical fiber cable is a cable containing one or more optical fibers. The optical fiber elements are typically individually coated with plastic layers and contained in a protective tube suitable for the environment where the cable will be deployed.

In practical fibers, the cladding is usually coated with a tough resin buffer layer, which may be further surrounded by a jacket layer, usually plastic.

Nuclear fusion

In physics and nuclear chemistry, nuclear fusion is the process by which multiple like-charged atomic nuclei join together to form a heavier nucleus. It is accompanied by the release or absorption of energy. Iron and nickel nuclei have the largest binding energies per nucleon of all nuclei.

Optical fiber

An Optical fiber (or fibre) is a glass or plastic fiber that carries light along its length. Fiber optics is the overlap of applied science and engineering concerned with the design and application of Optical fiber s. Optical fiber s are widely used in fiber-optic communications, which permits transmission over longer distances and at higher bandwidths (data rates) than other forms of communications.

Reflection

Reflection is the change in direction of a wave front at an interface between two different media so that the wave front returns into the medium from which it originated. Common examples include the reflection of light, sound and water waves.

Law of reflection: Angle of incidence = Angle of reflection

Reflections may occur in a number of wave and particle phenomena; these include acoustic, seismic waves in geologic structures, surface waves in bodies of water, and various electromagnetic waves, most usefully from VHF and higher radar frequencies, progressing upward through centimeter to millimeter-wavelength radar and the various light frequencies and (with special 'grazing' mirrors, to X-ray frequencies and beyond to gamma rays.

Total internal reflection

Total internal reflection is an optical phenomenon that occurs when a ray of light strikes a medium boundary at an angle larger than the critical angle with respect to the normal to the surface. If the refractive index is lower on the other side of the boundary no light can pass through, so effectively all of the light is reflected. The critical angle is the angle of incidence above which the total internal reflection occurs.

Reproduction

Reproduction is the biological process by which new individual organisms are produced. Reproduction is a fundamental feature of all known life; each individual organism exists as the result of reproduction. The known methods of reproduction are broadly grouped into two main types: sexual and asexual.

Physicist

A Physicist is a scientist who studies or practices physics. Physicist s study a wide range of physical phenomena in many branches of physics spanning all length scales: from sub-atomic particles of which all ordinary matter is made (particle physics) to the behavior of the material Universe as a whole (cosmology.)

Most material a student encounters in the undergraduate physics curriculum is based on discoveries and insights of a century or more in the past.

Rutherford

The Rutherford is an obsolete unit of radioactivity, defined as the activity of a quantity of radioactive material in which one million nuclei decay per second. It is therefore equivalent to one megabecquerel. It was named after Ernest Rutherford It is not an SI unit.

Solvay Conferences

The International Solvay Institutes for Physics and Chemistry, located in Brussels, were founded by the Belgian industrialist Ernest Solvay in 1912, following the historic invitation-only 1911 Conseil Solvay, the first world physics conference. The Institutes coordinate conferences, workshops, seminars, and colloquia.

Following the initial success of 1911, the Solvay Conferences have been devoted to outstanding preeminent open problems in both physics and chemistry.

Thermal neutron

A thermal neutron is a free neutron that is Boltzmann distributed with kT = 0.024 eV (4.0×10^{-21} J) at room temperature. This gives characteristic (not average, or median) speed of 2.2 km/s. The name 'thermal' comes from their energy being that of the room temperature gas or material they are permeating.

Nuclear force

The nuclear force is the force between two or more nucleons. It is responsible for binding of protons and neutrons into atomic nuclei. To a large extent, this force can be understood in terms of the exchange of virtual light mesons, such as the pions.

Nucleus

The nucleus of an atom is the very dense region, consisting of nucleons, at the center of an atom. Although the size of the nucleus varies considerably according to the mass of the atom, the size of the entire atom is comparatively constant. Almost all of the mass in an atom is made up from the protons and neutrons in the nucleus with a very small contribution from the orbiting electrons.

Theory of relativity

The theory of relativity refers specifically to two theories of Albert Einstein: special relativity and general relativity. However, 'relativity' can also refer to Galilean relativity.

The term 'theory of relativity' was coined by Max Planck in 1908 to emphasize how special relativity uses the principle of relativity.

Nuclear physics	Nuclear physics is the field of physics that studies the building blocks and interactions of atomic nuclei. It must not be confused with atomic physics, that studies the combined system of the nucleus and its arrangement of electrons, even if both terms are sometimes used synonymously in standard English. Particle physics is a field that has evolved out of Nuclear physics and for this reason has been included under the same term in earlier times.
Nucleus	The nucleus of an atom is the very dense region, consisting of nucleons, at the center of an atom. Although the size of the nucleus varies considerably according to the mass of the atom, the size of the entire atom is comparatively constant. Almost all of the mass in an atom is made up from the protons and neutrons in the nucleus with a very small contribution from the orbiting electrons.
Radioactivity	Radioactivity can be used in life sciences as a radiolabel to easily visualise components or target molecules in a biological system. Radionuclei are synthesised in particle accelerators and have short half-lives, giving them high maximum theoretical specific activities. This lowers the detection time compared to radionuclei with longer half-lives, such as carbon-14.
Sensor	A sensor is a device that measures a physical quantity and converts it into a signal which can be read by an observer or by an instrument. For example, a mercury thermometer converts the measured temperature into expansion and contraction of a liquid which can be read on a calibrated glass tube. A thermocouple converts temperature to an output voltage which can be read by a voltmeter.
Atomic mass	The Atomic mass (m_a) is the mass of an atom, most often expressed in unified Atomic mass units. The Atomic mass may be considered to be the total mass of protons, neutrons and electrons in a single atom (when the atom is motionless.) The Atomic mass is sometimes incorrectly used as a synonym of relative Atomic mass average Atomic mass and atomic weight; however, these differ subtly from the Atomic mass
Electron	An Electron is a subatomic particle that carries a negative electric charge. It has no known substructure and is believed to be a point particle. An Electron has a mass that is approximately 1836 times less than that of the proton.
Mass	Mass is a concept used in the physical sciences to explain a number of observable behaviours, and in everyday usage, it is common to identify Mass with those resulting behaviors. In particular, Mass is commonly identified with weight. But according to our modern scientific understanding, the weight of an object results from the interaction of its Mass with a gravitational field, so while Mass is part of the explanation of weight, it is not the complete explanation.
Thermal neutron	A thermal neutron is a free neutron that is Boltzmann distributed with kT = 0.024 eV (4.0×10^{-21} J) at room temperature. This gives characteristic (not average, or median) speed of 2.2 km/s. The name 'thermal' comes from their energy being that of the room temperature gas or material they are permeating.
Proton	The Proton is a subatomic particle with an electric charge of +1 elementary charge. It is found in the nucleus of each atom but is also stable by itself and has a second identity as the hydrogen ion, $^{1}H^{+}$. It is composed of three even more fundamental particles comprising two up quarks and one down quark.

Isotopes	Isotopes are different types of atoms (nuclides) of the same chemical element, each having a different atomic mass (mass number.) Isotopes of an element have nuclei with the same number of protons (the same atomic number) but different numbers of neutrons. Therefore, Isotopes of the same element have different mass numbers (number of nucleons.)
Nuclear power	Nuclear power is any nuclear technology designed to extract usable energy from atomic nuclei via controlled nuclear reactions. The most common method today is through nuclear fission, though other methods include nuclear fusion and radioactive decay. All utility-scale reactors heat water to produce steam, which is then converted into mechanical work for the purpose of generating electricity or propulsion.
Nuclear reaction	In nuclear physics, a nuclear reaction is the process in which two nuclei or nuclear particles collide to produce products different from the initial particles. In principle a reaction can involve more than two particles colliding, but because the probability of three or more nuclei to meet at the same time at the same place is much less than for two nuclei, such an event is exceptionally rare. While the transformation is spontaneous in the case of radioactive decay, it is initiated by a particle in the case of a nuclear reaction.
Nucleon	In physics a nucleon is a collective name for two baryons: the neutron and the proton. They are constituents of the atomic nucleus and until the 1960s were thought to be elementary particles. In those days their interactions defined strong interactions.
Power station	A power station is an industrial facility for the generation of electric power. Power plant is also used to refer to the engine in ships, aircraft and other large vehicles. Some prefer to use the term energy center because it more accurately describes what the plants do, which is the conversion of other forms of energy, like chemical energy, gravitational potential energy or heat energy into electrical energy.
Radioactive	Radioactive decay is the process in which an unstable atomic nucleus loses energy by emitting ionizing particles and radiation. This decay, or loss of energy, results in an atom of one type, called the parent nuclide transforming to an atom of a different type, called the daughter nuclide. For example: a carbon-14 atom emits radiation and transforms to a nitrogen-14 atom.
Radioactive decay	Radioactive decay is the process in which an unstable atomic nucleus loses energy by emitting ionizing particles and radiation. This decay, or loss of energy, results in an atom of one type, called the parent nuclide transforming to an atom of a different type, called the daughter nuclide. For example: a carbon-14 atom emits radiation and transforms to a nitrogen-14 atom.
Uranium	Uranium is a silvery-gray metallic chemical element in the actinide series of the periodic table that has the symbol U and atomic number 92. It has 92 protons and 92 electrons, 6 of them valence electrons. It can have between 141 and 146 neutrons, with 146 and 143 in its most common isotopes.
Electrostatic	Electrostatic s is the branch of science that deals with the phenomena arising from stationary or slowly moving electric charges. Since classical antiquity it was known that some materials such as amber attract light particles after rubbing. The Greek word for amber, Î®λεκτρον , was the source of the word 'electricity'.

Force

In physics, a Force is any external agent that causes a change in the motion of a free body, or that causes stress in a fixed body. It can also be described by intuitive concepts such as a push or pull that can cause an object with mass to change its velocity , i.e., to accelerate, or which can cause a flexible object to deform. Force has both magnitude and direction, making it a vector quantity.

Nuclear force

The nuclear force is the force between two or more nucleons. It is responsible for binding of protons and neutrons into atomic nuclei. To a large extent, this force can be understood in terms of the exchange of virtual light mesons, such as the pions.

Strong

In astronomy in particular, for the description of celestial sources of radiation, Strong is understood to refer to intensity when observed from the Earth or another particular body, relative to other sources relevant in the same observational context. In the optical (visual) range, 'brightness' is synonymous, outside it 'Strong' is standard. Absolute strength, taking into account distance effects, and generally integrating over frequency and over all directions, is termed luminosity, and measured in units of energy per unit time.

Strong interaction

In particle physics, the strong interaction, or strong force, or color force, holds quarks and gluons together to form protons and neutrons. The strong interaction is one of the four fundamental interactions, along with gravitation, the electromagnetic force and the weak interaction. The word strong is used since the strong interaction is the most powerful of the four fundamental forces; its typical field strength is 100 times the strength of the electromagnetic force, some 10^{13} times as great as that of the weak force, and about 10^{38} times that of gravitation.

Weak force

The weak force is one of the four fundamental interactions of nature. In the Standard Model of particle physics, it is due to the exchange of the heavy W and Z bosons. Its most familiar effect is beta decay and the associated radioactivity.

Alpha decay

Alpha decay is a type of radioactive decay in which an atomic nucleus emits an alpha particle (two protons and two neutrons bound together into a particle identical to a helium nucleus) and transforms (or 'decays') into an atom with a mass number 4 less and atomic number 2 less. For example:

(The second form is preferred because the first form appears electrically unbalanced. Fundamentally, the recoiling nucleus is very quickly stripped of the two extra electrons which give it an unbalanced charge.

Beta decay

In nuclear physics, Beta decay is a type of radioactive decay in which a beta particle (an electron or a positron) is emitted. In the case of electron emission, it is referred to as beta minus (β^-), while in the case of a positron emission as beta plus (β^+.) Kinetic energy of beta particles has continuous spectrum ranging from 0 to maximal available energy (Q), which depends on parent and daughter nuclear states participating in the decay.

Electric field

In physics, the space surrounding an electric charge or in the presence of a time-varying magnetic field has a property called an Electric field This Electric field exerts a force on other electrically charged objects. The concept of an Electric field was introduced by Michael Faraday.

Gamma rays

Gamma rays (denoted as γ) are electromagnetic radiation of high energy. They are produced by sub-atomic particle interactions, such as electron-positron annihilation, neutral pion decay, radioactive decay, fusion, fission or inverse Compton scattering in astrophysical processes. Gamma rays typically have frequencies above 10^{19} Hz and therefore energies above 100 keV and wavelength less than 10 picometers, often smaller than an atom.

Magnetic field

Magnetic field s surround magnetic materials and electric currents and are detected by the force they exert on other magnetic materials and moving electric charges. The Magnetic field at any given point is specified by both a direction and a magnitude (or strength); as such it is a vector field.

For the physics of magnetic materials, see magnetism and magnet, more specifically ferromagnetism, paramagnetism, and diamagnetism.

Photon

In physics, the photon is the elementary particle responsible for electromagnetic phenomena. It is the carrier of electromagnetic radiation of all wavelengths, including in decreasing order of energy, gamma rays, X-rays, ultraviolet light, visible light, infrared light, microwaves, and radio waves. The photon differs from many other elementary particles, such as the electron and the quark, in that it has zero rest mass; therefore, it travels at the speed of light, c.

Radiation

Radiation, as in physics, is energy in the form of waves or moving subatomic particles emitted by an atom or other body as it changes from a higher energy state to a lower energy state. Radiation can be classified as ionizing or non-ionizing radiation, depending on its effect on atomic matter. The most common use of the word 'radiation' refers to ionizing radiation.

Radionuclide

A radionuclide is an atom with an unstable nucleus, which is a nucleus characterized by excess energy which is available to be imparted either to a newly-created radiation particle within the nucleus, or else to an atomic electron . The radionuclide, in this process, undergoes radioactive decay, and emits a gamma ra and/or subatomic particles. These particles constitute ionizing radiation.

Wave

A Wave is a disturbance that propagates through space and time, usually with transference of energy. A mechanical Wave is a Wave that propagates or travels through a medium due to the restoring forces it produces upon deformation. There also exist Wave s capable of traveling through a vacuum, including electromagnetic radiation and probably gravitational radiation.

Waves

Waves is a three-part novel by Ogan Gurel published in 2009. A 21st century version of Faust, the novel explores good and evil in both individual and global settings focusing around a hypothetical technology that has both medical and military applications. The protagonist, Tomas Twarok, is a contemplative and idealistic doctor-turned-entrepreneur who strikes a deal with his college friend, Maximilian Iblis, a ruthless hedge fund manager.

Plutonium

Plutonium is a rare radioactive, metallic chemical element. The most significant isotope of plutonium is ^{239}Pu, with a half-life of 24,100 years; this isotope is fissile and is used in most modern nuclear weapons. Plutonium-239 can be synthesized from natural uranium.

Lead

Lead is a main-group element with symbol Pb and atomic number 82. Lead is a soft, malleable poor metal, also considered to be one of the heavy metals. Lead has a bluish-white color when freshly cut, but tarnishes to a dull grayish color when exposed to air.

Molecule

A Molecule is defined as a sufficiently stable, electrically neutral group of at least two atoms in a definite arrangement held together by very strong (covalent) chemical bonds. Molecule s are distinguished from polyatomic ions in this strict sense. In organic chemistry and biochemistry, the term Molecule is used less strictly and also is applied to charged organic Molecule s and bio Molecule s.

Strontium	Strontium is a chemical element with the symbol Sr and the atomic number 38. An alkaline earth metal, strontium is a soft silver-white or yellowish metallic element that is highly reactive chemically. The metal turns yellow when exposed to air.
Chain	A Chain is a series of connected links A Chain may consist of two or more links.
Earth	Earth is the third planet from the Sun. It is the fifth largest of the eight planets in the solar system, and the largest of the terrestrial planets (non-gas planets) in the Solar System in terms of diameter, mass and density. It is also referred to as the World, the Blue Planet, and Terra.
Electric charge	Electric charge is a fundamental conserved property of some subatomic particles, which determines their electromagnetic interaction. Electrically charged matter is influenced by, and produces, electromagnetic fields. The interaction between a moving charge and an electromagnetic field is the source of the electromagnetic force, which is one of the four fundamental forces.
Internal energy	In thermodynamics, the Internal energy of a thermodynamic system denoted by U is the total of the kinetic energy due to the motion of molecules (translational, rotational, vibrational) and the potential energy associated with the vibrational and electric energy of atoms within molecules or crystals. It includes the energy in all of the chemical bonds, and the energy of the free, conduction electrons in metals. One can also calculate the Internal energy of electromagnetic or blackbody radiation.
Kinetic energy	The Kinetic energy of an object is the extra energy which it possesses due to its motion. It is defined as the work needed to accelerate a body of a given mass from rest to its current velocity. Having gained this energy during its acceleration, the body maintains this Kinetic energy unless its speed changes.
Nuclear medicine	Nuclear medicine is a branch of medicine and medical imaging that uses the nuclear properties of matter in diagnosis and therapy. More specifically, nuclear medicine is a part of molecular imaging because it produces images that reflect biological processes that take place at the cellular and subcellular level. Para-sagittal MRI of the head in a patient with benign familial macrocephaly.
Potential energy	Potential energy can be thought of as energy stored within a physical system. It is called potential energy because it has the potential to be converted into other forms of energy, such as kinetic energy, and to do work in the process. The standard unit of measure for potential energy is the joule, the same as for work, or energy in general.
Solid	The solid state of matter is characterized by a distinct structural rigidity and virtual resistance to deformation (i.e. changes of shape and/or volume.) Most solid s have high values both of Young's modulus and of the shear modulus of elasticity. This contrasts with most liquids or fluids, which have a low shear modulus, and typically exhibit the capacity for macroscopic viscous flow.
Spacecraft	A spacecraft is a vehicle or machine designed for spaceflight. On a sub-orbital spaceflight, a spacecraft enters outer space but then returns to the planetary surface without making a complete orbit. For an orbital spaceflight, a spacecraft enters a closed orbit around the planetary body.

Transmitter	A transmitter is an electronic device which, usually with the aid of an antenna, propagates an electromagnetic signal such as radio, television, or other telecommunications. In other applications signals can also be transmitted using an analog 0/4-20 mA current loop signal. WDET-FM transmitter Generally and in communication and information processing, a transmitter is any object which sends information to an observer.
Nuclear reactor	A Nuclear reactor is a device in which nuclear chain reactions are initiated, controlled, and sustained at a steady rate. The most significant use of Nuclear reactor s is as an energy source for the generation of electrical power and for the power in some ships This is usually accomplished by methods that involve using heat from the nuclear reaction to power steam turbines.
X-radiation	X-radiation is a form of electromagnetic radiation. X-rays have a wavelength in the range of 10 to 0.01 nanometers, corresponding to frequencies in the range 30 petahertz to 30 exahertz and energies in the range 120 eV to 120 keV. They are longer than gamma rays but shorter than UV rays.
Elementary particle	In particle physics, an Elementary particle or fundamental particle is a particle not known to have substructure; that is, it is not known to be made up of smaller particles. If an Elementary particle truly has no substructure, then it is one of the basic building blocks of the universe from which all other particles are made. In the Standard Model, the quarks, leptons, and gauge bosons are Elementary particle s.
Binding energy	Binding energy is the mechanical energy required to disassemble a whole into separate parts. A bound system has typically a lower potential energy than its constituent parts; this is what keeps the system together. The usual convention is that this corresponds to a positive Binding energy
Nuclear energy	Nuclear Energy is released by the splitting or merging together of the nuclei of ato. The conversion of nuclear mass to energy is consistent with the mass-energy equivalence formula $\Delta E = \Delta m.c^2$, in which ΔE = energy release, Δm = mass defect, and c = the speed of light in a vacuum. Nuclear energy was first discovered by French physicist Henri Becquerel in 1896, when he found that photographic plates stored in the dark near uranium were blackened like X-ray plates, which had been just recently discovered at the time 1895.
Matter	The term Matter traditionally refers to the substance that objects are made of. One common way to identify this 'substance' is through its physical properties; a common definition of Matter is anything that has mass and occupies a volume. However, this definition has to be revised in light of quantum mechanics, where the concept of 'having mass', and 'occupying space' are not as well-defined as in everyday life.
Nuclear fission	Nuclear fission is the splitting of the nucleus of an atom into parts often producing free neutrons and other smaller nuclei, which may eventually produce photons. Fission of heavy elements is an exothermic reaction which can release large amounts of energy both as electromagnetic radiation and as kinetic energy of the fragments. Fission is a form of elemental transmutation because the resulting fragments are not the same element as the original atom.
Nuclear fusion	In physics and nuclear chemistry, nuclear fusion is the process by which multiple like-charged atomic nuclei join together to form a heavier nucleus. It is accompanied by the release or absorption of energy. Iron and nickel nuclei have the largest binding energies per nucleon of all nuclei.

Theory of relativity

The theory of relativity refers specifically to two theories of Albert Einstein: special relativity and general relativity. However, 'relativity' can also refer to Galilean relativity.

The term 'theory of relativity' was coined by Max Planck in 1908 to emphasize how special relativity uses the principle of relativity.

Critical mass

A Critical mass is the smallest amount of fissile material needed for a sustained nuclear chain reaction. The Critical mass of a fissionable material depends upon its nuclear properties (e.g. the nuclear fission cross-section), its density, its shape, its enrichment, its purity, its temperature, and its surroundings.

The term critical refers to an equilibrium fission reaction (steady-state or continuous chain reaction); this is where there is no increase or decrease in power, temperature, or neutron population.

Collision

A Collision is an isolated event in which two or more moving bodies (colliding bodies) exert relatively strong forces on each other for a relatively short time. Deflection happens when an object hits a plane surface

Collision s involve forces (there is a change in velocity.) Collision s can be elastic, meaning they conserve energy and momentum, inelastic, meaning they conserve momentum but not energy, or totally inelastic (or plastic), meaning they conserve momentum and the two objects stick together.

Thermal

A thermal column (or thermal is a column of rising air in the lower altitudes of the Earth's atmosphere. thermal s are created by the uneven heating of the Earth's surface from solar radiation, and an example of convection. The Sun warms the ground, which in turn warms the air directly above it.

Rods

Rod may mean:

· Rod, a straight and slender stick; a wand; a cylinder; hence, any slender bar
· Rod cell, a cell found in the retina that is sensitive to light/dark
· Rod, an Imperial unit of length observational artifacts produced by rapidly flying animals
· Rod, a Slavic god
· Rods
· Railway Operating Division
· Fishing rod
· Lightning rod
· Connecting rod, in an internal combustion engine
· Divining rod, two rods believed by some to find water in a practice known as dowsing
· Birch rod, made out of twigs from birch or other trees for corporal punishment
· Switch, a piece of wood as used as a staff or for corporal punishment, or a bundle of such switches
· Ring of Death, a common malfunction of the Xbox 360

Rod may also be:

· Eubacterium with the shape of a rod; a bacillus in the loose sense
· Penis
· Gun

Often as an abbreviation of such names as Roderick and Rodney:

· Rod Stewart, rock musician
· Rod Laver, Australian former tennis player
· Rod Steiger, American actor
· Rod Flanders, a fictional character on The Simpsons
· Rod From Avenue Q, a character in the stage musical Avenue Q
· Rod Serling, Creator of the 1960's show The Twilight Zone
· Rod Quantock, Australian stand-up comedian
· Rod Roddy, Game show announcer, best known for The Price is Right
· Darren Hampton, Tesco Meir Park

· Edouard Rod, French-Swiss novelist, 1857-1910
· Johnny Rod, American bass guitar player .

Periodic table

The periodic table of the chemical elements is a tabular method of displaying the chemical elements. Although precursors to this table exist, its invention is generally credited to Russian chemist Dmitri Mendeleev in 1869. Mendeleev intended the table to illustrate recurring trends in the properties of the elements.

Solar

Solar power uses Solar Radiation emitted from our sun. Solar power, a renewable energy source, has been used in many traditional technologies for centuries, and is in widespread use where other power supplies are absent, such as in remote locations and in space.

Solar energy

Solar energy is the light and radiant heat from the Sun that influences Earth's climate and weather and sustains life. Solar power is the rate of solar energy at a point in time; it is sometimes used as a synonym for solar energy or more specifically to refer to electricity generated from solar radiation. Since ancient times solar energy has been harnessed for human use through a range of technologies.

Star

A star is a massive, luminous ball of plasma that is held together by gravity. The nearest star to Earth is the Sun, which is the source of most of the energy on Earth. Other star s are visible in the night sky, when they are not outshone by the Sun.

Supernova

A supernova is a stellar explosion. They are extremely luminous and cause a burst of radiation that often briefly outshines an entire galaxy before fading from view over several weeks or months. During this short interval, a supernova can radiate as much energy as the Sun could emit over its life span.

Temperature

Temperature is a physical property of a system that underlies the common notions of hot and cold; something that is hotter generally has the greater temperature. Specifically, temperature is a property of matter. Temperature is one of the principal parameters of thermodynamics.

Antiparticle

· Particle accelerator
· Penning trap

Antiparticle s

· Positron
· Antiproton
· Antineutron

Uses

· Positron emission tomography
· Fuel
· Weaponry

Bodies

· ALPHA Collaboration
· ATHENA
· ATRAP
· CERN

People

Illustration of electric charge as well as general size of particles (left) and Antiparticle s (right.) From top to bottom; electron/positron, proton/antiproton, neutron/antineutron.

· Paul Dirac
· Carl David Anderson
· Andrei Sakharov

Corresponding to most kinds of particles, there is an associated Antiparticle with the same mass and opposite electric charge. For example, the Antiparticle of the electron is the positively charged antielectron, or positron, which is produced naturally in certain types of radioactive decay.

Big Bang

The Big Bang is the cosmological model of the initial conditions and subsequent development of the universe that is supported by the most comprehensive and accurate explanations from current scientific evidence and observation. As used by cosmologists, the term Big Bang generally refers to the idea that the universe has expanded from a primordial hot and dense initial condition at some finite time in the past (currently estimated to have been approximately 13.7 billion years ago), and continues to expand to this day.

Georges Lemaître proposed what became known as the Big Bang theory of the origin of the Universe, although he called it his 'hypothesis of the primeval atom'.

Positron

The positron or antielectron is the antiparticle or the antimatter counterpart of the electron. The positron has an electric charge of +1, a spin of 1/2, and the same mass as an electron. When a low-energy positron collides with a low-energy electron, annihilation occurs, resulting in the production of two gamma ray photons.

Plasma

In physics and chemistry, plasma is an ionized gas, in which a certain proportion of electrons are free rather than being bound to an atom or molecule. The ability of the positive and negative charges to move somewhat independently makes the plasma electrically conductive so that it responds strongly to electromagnetic fields. Plasma therefore has properties quite unlike those of solids, liquids or gases and is considered to be a distinct state of matter.

Pulsed power	Pulsed power is the term used to describe the science and technology of accumulating energy over a relatively long period of time and releasing it very quickly thus increasing the instantaneous power. Steady accumulation of energy followed by its rapid release can result in the delivery of a larger amount of instantaneous power over a shorter period of time. Energy is typically stored within electrostatic fields, magnetic fields, as mechanical energy, or as chemical energy.
Tritium	Tritium is a radioactive isotope of hydrogen. The nucleus of tritium contains one proton and two neutrons, whereas the nucleus of protium contains no neutrons and one proton. While Tritium has several different experimentally-determined values of its half-life, the NIST recommends 4500±8 days .
Z machine	The Z machine is the largest X-ray generator in the world and is designed to test materials in conditions of extreme temperature and pressure. Operated by Sandia National Laboratories, it gathers data to aid in computer modeling of nuclear weapons. The Z machine is located at Sandia's main site in Albuquerque, New Mexico.
Cold	Cold refers to the condition or perception of having low temperature, it is the absence of heat or warmth. Many things are associated with Cold such as ice and the color blue. Fluids used to cool objects are commonly called coolants.
Atomic physics	Atomic physics (or atom physics) is the field of physics that studies atoms as an isolated system of electrons and an atomic nucleus. It is primarily concerned with the arrangement of electrons around the nucleus and the processes by which these arrangements change. This includes ions as well as neutral atoms and, unless otherwise stated, for the purposes of this discussion it should be assumed that the term atom includes ions.
Nitrogen	Nitrogen is a chemical element that has the symbol N and atomic number 7 and atomic weight 14.0067. Elemental nitrogen is a colorless, odorless, tasteless and mostly inert diatomic gas at standard conditions, constituting 80% by volume of Earth's atmosphere. Many industrially important compounds, such as ammonia, nitric acid, organic nitrates, and cyanides, contain nitrogen.
Rutherford	The Rutherford is an obsolete unit of radioactivity, defined as the activity of a quantity of radioactive material in which one million nuclei decay per second. It is therefore equivalent to one megabecquerel. It was named after Ernest Rutherford It is not an SI unit.
Enrico Fermi	Enrico Fermi (29 September 1901 - 28 November 1954) was an Italian physicist most noted for his work on the development of the first nuclear reactor, and for his contributions to the development of quantum theory, nuclear and particle physics, and statistical mechanics. Fermi was awarded the Nobel Prize in Physics in 1938 for his work on induced radioactivity and is today regarded as one of the top scientists of the 20th century. He is acknowledged as a unique physicist who was highly accomplished in both theory and experiment.
Otto Hahn	Otto Hahn was a German chemist who received the 1944 Nobel Prize in Chemistry for discovering nuclear fission. He is considered a pioneer of radioactivity and radiochemistry. Glenn T.

Antimatter

- Particle accelerator
- Penning trap

Antiparticles

- Positron
- Antiproton
- Antineutron

Uses

- Positron emission tomography
- Fuel
- Weaponry

Bodies

- ALPHA Collaboration
- ATHENA
- ATRAP
- CERN

People

- Paul Dirac
- Carl David Anderson
- Andrei Sakharov

In particle physics, Antimatter is the extension of the concept of the antiparticle to matter, where Antimatter is composed of antiparticles in the same way that normal matter is composed of particles. For example, an antielectron (a positron, an electron with a positive charge) and an antiproton (a proton with a negative charge) could form an antihydrogen atom in the same way that an electron and a proton form a normal matter hydrogen atom. Furthermore, mixing matter and Antimatter would lead to the annihilation of both in the same way that mixing antiparticles and particles does, thus giving rise to high-energy photons (gamma rays) or other particle-antiparticle pairs.

Big Bang

The Big Bang is the cosmological model of the initial conditions and subsequent development of the universe that is supported by the most comprehensive and accurate explanations from current scientific evidence and observation. As used by cosmologists, the term Big Bang generally refers to the idea that the universe has expanded from a primordial hot and dense initial condition at some finite time in the past (currently estimated to have been approximately 13.7 billion years ago), and continues to expand to this day.

Georges Lemaître proposed what became known as the Big Bang theory of the origin of the Universe, although he called it his 'hypothesis of the primeval atom'.

Cosmology

Cosmology is the study of the Universe in its totality, and by extension, humanity's place in it. Though the word Cosmology is recent ' href='/wiki/Christian_Wolff_(philosopher)'>Christian Wolff's Cosmologia Generalis), study of the universe has a long history involving science, philosophy, esotericism, and religion.

In recent times, physics and astrophysics have played a central role in shaping the understanding of the universe through scientific observation and experiment; or what is known as physical Cosmology shaped through both mathematics and observation in the analysis of the whole universe.

Elementary particle	In particle physics, an Elementary particle or fundamental particle is a particle not known to have substructure; that is, it is not known to be made up of smaller particles. If an Elementary particle truly has no substructure, then it is one of the basic building blocks of the universe from which all other particles are made. In the Standard Model, the quarks, leptons, and gauge bosons are Elementary particle s.
Gamma rays	Gamma rays (denoted as γ) are electromagnetic radiation of high energy. They are produced by sub-atomic particle interactions, such as electron-positron annihilation, neutral pion decay, radioactive decay, fusion, fission or inverse Compton scattering in astrophysical processes. Gamma rays typically have frequencies above 10^{19} Hz and therefore energies above 100 keV and wavelength less than 10 picometers, often smaller than an atom.
Particle physics	Particle physics is a branch of physics that studies the elementary constituents of matter and radiation, and the interactions between them. It is also called high energy physics, because many elementary particles do not occur under normal circumstances in nature, but can be created and detected during energetic collisions of other particles, as is done in particle accelerators. Research in this area has produced a long list of particles.
Positron	The positron or antielectron is the antiparticle or the antimatter counterpart of the electron. The positron has an electric charge of +1, a spin of 1/2, and the same mass as an electron. When a low-energy positron collides with a low-energy electron, annihilation occurs, resulting in the production of two gamma ray photons.
Positron emission	Positron emission is a type of beta decay, sometimes referred to as 'beta plus'. In beta plus decay, a proton is converted, via the weak force, to a neutron, a positron, and a neutrino. Isotopes which undergo this decay and thereby emit positrons include carbon-11, potassium-40, nitrogen-13, oxygen-15, fluorine-18, and iodine-121.
Positron emission tomography	Positron emission tomography is a nuclear medicine imaging technique which produces a three-dimensional image or map of functional processes in the body. The system detects pairs of gamma rays emitted indirectly by a positron-emitting radionuclide, which is introduced into the body on a biologically active molecule. Images of tracer concentration in 3-dimensional space within the body are then reconstructed by computer analysis.
Radioactive	Radioactive decay is the process in which an unstable atomic nucleus loses energy by emitting ionizing particles and radiation. This decay, or loss of energy, results in an atom of one type, called the parent nuclide transforming to an atom of a different type, called the daughter nuclide. For example: a carbon-14 atom emits radiation and transforms to a nitrogen-14 atom.
Radioactive decay	Radioactive decay is the process in which an unstable atomic nucleus loses energy by emitting ionizing particles and radiation. This decay, or loss of energy, results in an atom of one type, called the parent nuclide transforming to an atom of a different type, called the daughter nuclide. For example: a carbon-14 atom emits radiation and transforms to a nitrogen-14 atom.
Radionuclide	A radionuclide is an atom with an unstable nucleus, which is a nucleus characterized by excess energy which is available to be imparted either to a newly-created radiation particle within the nucleus, or else to an atomic electron . The radionuclide, in this process, undergoes radioactive decay, and emits a gamma ra and/or subatomic particles. These particles constitute ionizing radiation.

Theory of relativity

The theory of relativity refers specifically to two theories of Albert Einstein: special relativity and general relativity. However, 'relativity' can also refer to Galilean relativity.

The term 'theory of relativity' was coined by Max Planck in 1908 to emphasize how special relativity uses the principle of relativity.

Special relativity

Special relativity is the physical theory of measurement in inertial frames of reference proposed in 1905 by Albert Einstein in the paper 'On the Electrodynamics of Moving Bodies'. It generalizes Galileo's principle of relativity - that all uniform motion is relative, and that there is no absolute and well-defined state of rest - from mechanics to all the laws of physics, including both the laws of mechanics and of electrodynamics, whatever they may be. In addition, special relativity incorporates the principle that the speed of light is the same for all inertial observers regardless of the state of motion of the source.

Standard Model

The Standard Model of particle physics is a theory that describes three of the four known fundamental interactions together with the elementary particles that take part in these interactions. These particles make up all matter in the universe except for the dark matter. The standard model is a non-abelian gauge theory of the electroweak and strong interactions with the symmetry group S×S×.

Sublimation

Sublimation of an element or compound is a transition from the solid to gas phase so rapidly that the liquid phase cannot be observed. Sublimation is a phase transition that occurs at temperatures and pressures below the triple point.

Mechanics

Mechanics is the branch of physics concerned with the behaviour of physical bodies when subjected to forces or displacements, and the subsequent effect of the bodies on their environment. The discipline has its roots in several ancient civilizations During the early modern period, scientists such as Galileo, Kepler, and especially Newton, laid the foundation for what is now known as classical Mechanics

Universal gravitation

Newton's law of universal gravitation is a physical law describing the gravitational attraction between bodies with mass. It is a part of classical mechanics and was first formulated in Newton's work Philosophiae Naturalis Principia Mathematica, first published on July 5, 1687. In modern language it states the following:

Every point mass attracts every other point mass by a force pointing along the line intersecting both points.

Velocity

In physics, velocity is defined as the rate of change of position. It is a vector physical quantity; both speed and direction are required to define it. In the SI system, it is measured in meters per second: or ms^{-1}.

Conservation law

In physics, a Conservation law states that a particular measurable property of an isolated physical system does not change as the system evolves.

One particularly important physical result concerning Conservation law s is Noether's Theorem, which states that there is a one-to-one correspondence between Conservation law s and differentiable symmetries of physical systems. For example, the conservation of energy follows from the time-invariance of physical systems, and the fact that physical systems behave the same regardless of how they are oriented in space gives rise to the conservation of angular momentum.

Time dilation

Time dilation is the phenomenon whereby an observer finds that another's clock, which is physically identical to their own, is ticking at a slower rate as measured by their own clock. This is often interpreted as time 'slowing down' for the other clock, but that is only true in the context of the observer's frame of reference. Locally, time always passes at the same rate.

Pion

In particle physics, pion is the collective name for three subatomic particles: π^0, π^+ and π^-. Pions are the lightest mesons and play an important role in explaining low-energy properties of the strong nuclear force. Feynman diagram of a common pion decay.

Pions have zero spin and are composed of first-generation quarks.

Cosmic ray

Cosmic ray s are energetic particles originating from outer space that impinge on Earth's atmosphere. Almost 90% of all the incoming Cosmic ray particles are protons, almost 10% are helium nuclei (alpha particles), and slightly under 1% are heavier elements and electrons (beta minus particles.) The term ray is a misnomer, as cosmic particles arrive individually, not in the form of a ray or beam of particles.

Length

Length is the long dimension of any object. The Length of a thing is the distance between its ends, its linear extent as measured from end to end. This may be distinguished from height, which is vertical extent, and width or breadth, which are the distance from side to side, measuring across the object at right angles to the Length

Collision

A Collision is an isolated event in which two or more moving bodies (colliding bodies) exert relatively strong forces on each other for a relatively short time. Deflection happens when an object hits a plane surface

Collision s involve forces (there is a change in velocity.) Collision s can be elastic, meaning they conserve energy and momentum, inelastic, meaning they conserve momentum but not energy, or totally inelastic (or plastic), meaning they conserve momentum and the two objects stick together.

Energy

In physics, Energy is a scalar physical quantity that describes the amount of work that can be performed by a force, an attribute of objects and systems that is subject to a conservation law. Different forms of Energy include kinetic, potential, thermal, gravitational, sound, light, elastic, and electromagnetic Energy The forms of Energy are often named after a related force.

Mass

Mass is a concept used in the physical sciences to explain a number of observable behaviours, and in everyday usage, it is common to identify Mass with those resulting behaviors. In particular, Mass is commonly identified with weight. But according to our modern scientific understanding, the weight of an object results from the interaction of its Mass with a gravitational field, so while Mass is part of the explanation of weight, it is not the complete explanation.

Rest energy

The rest energy E or rest mass-energy of a particle is its energy when it is at rest relative to a given inertial reference frame. This type of energy can immediately change into Potential Energy and into Kinetic Energy. It is defined by the math formula as follows:

$$E = m_0 c^2,$$

where m_0 is the rest mass of the particle and c is the speed of light in a vacuum.

Inelastic collision

An Inelastic collision is a collision in which kinetic energy is not conserved

In collisions of macroscopic bodies, some kinetic energy is turned into vibrational energy of the atoms, causing a heating effect, and the bodies are deformed.

The molecules of a gas or liquid rarely experience perfectly elastic collisions because kinetic energy is exchanged between the molecules' translational motion and their internal degrees of freedom with each collision.

Kinetic energy

The Kinetic energy of an object is the extra energy which it possesses due to its motion. It is defined as the work needed to accelerate a body of a given mass from rest to its current velocity. Having gained this energy during its acceleration, the body maintains this Kinetic energy unless its speed changes.

Quantum

In physics, a quantum is an indivisible entity of a quantity that has the same units as the Planck constant and is related to both energy and momentum of elementary particles of matter and of photons and other bosons. The word comes from the Latin 'quantus,' for 'how much.' Behind this, one finds the fundamental notion that a physical property may be 'quantized', referred to as 'quantization'. This means that the magnitude can take on only certain discrete numerical values, rather than any value, at least within a range.

Quantum mechanics

Quantum mechanics is the study of mechanical systems whose dimensions are close to the atomic scale, such as molecules, atoms, electrons, protons and other subatomic particles. Quantum mechanics is a fundamental branch of physics with wide applications. Quantum theory generalizes classical mechanics to provide accurate descriptions for many previously unexplained phenomena such as black body radiation and stable electron orbits.

Low pressure area

A low pressure area is a region where the atmospheric pressure is lower in relation to the surrounding area. Low pressure systems form under areas of upper level divergence on the east side of upper troughs, or due to localized heating caused by greater insolation or active thunderstorm activity. Those that form due to organized thunderstorm activity over the water which acquire a well-defined circulation are called tropical cyclones.

Force

In physics, a Force is any external agent that causes a change in the motion of a free body, or that causes stress in a fixed body. It can also be described by intuitive concepts such as a push or pull that can cause an object with mass to change its velocity , i.e., to accelerate, or which can cause a flexible object to deform. Force has both magnitude and direction, making it a vector quantity.

Interaction

Interaction is a kind of action that occurs as two or more objects have an effect upon one another. The idea of a two-way effect is essential in the concept of Interaction as opposed to a one-way causal effect. A closely related term is interconnectivity, which deals with the Interaction s of Interaction s within systems: combinations of many simple Interaction s can lead to surprising emergent phenomena.

Strong

In astronomy in particular, for the description of celestial sources of radiation, Strong is understood to refer to intensity when observed from the Earth or another particular body, relative to other sources relevant in the same observational context. In the optical (visual) range, 'brightness' is synonymous, outside it 'Strong' is standard. Absolute strength, taking into account distance effects, and generally integrating over frequency and over all directions, is termed luminosity, and measured in units of energy per unit time.

Strong interaction

In particle physics, the strong interaction, or strong force, or color force, holds quarks and gluons together to form protons and neutrons. The strong interaction is one of the four fundamental interactions, along with gravitation, the electromagnetic force and the weak interaction. The word strong is used since the strong interaction is the most powerful of the four fundamental forces; its typical field strength is 100 times the strength of the electromagnetic force, some 10^{13} times as great as that of the weak force, and about 10^{38} times that of gravitation.

Weak force

The weak force is one of the four fundamental interactions of nature. In the Standard Model of particle physics, it is due to the exchange of the heavy W and Z bosons. Its most familiar effect is beta decay and the associated radioactivity.

Beta decay

In nuclear physics, Beta decay is a type of radioactive decay in which a beta particle (an electron or a positron) is emitted. In the case of electron emission, it is referred to as beta minus (β^-), while in the case of a positron emission as beta plus (β^+.) Kinetic energy of beta particles has continuous spectrum ranging from 0 to maximal available energy (Q), which depends on parent and daughter nuclear states participating in the decay.

Electromagnetism

Electromagnetism is the physics of the electromagnetic field, a field that exerts a force on particles with the property of electric charge and is reciprocally affected by the presence and motion of such particles.

A changing magnetic field produces an electric field (this is the phenomenon of electromagnetic induction, the basis of operation for electrical generators, induction motors, and transformers.) Similarly, a changing electric field generates a magnetic field.

Electroweak interaction

In particle physics, the Electroweak interaction is the unified description of two of the four fundamental interactions of nature: electromagnetism and the weak interaction. Although these two forces appear very different at everyday low energies, the theory models them as two different aspects of the same force. Above the unification energy, on the order of 100 GeV, they would merge into a single electroweak force.

Antiparticle

- Particle accelerator
- Penning trap

Antiparticle s

- Positron
- Antiproton
- Antineutron

Uses

- Positron emission tomography
- Fuel
- Weaponry

Bodies

· ALPHA Collaboration
· ATHENA
· ATRAP
· CERN

People

Illustration of electric charge as well as general size of particles (left) and Antiparticle s (right.) From top to bottom; electron/positron, proton/antiproton, neutron/antineutron.

· Paul Dirac
· Carl David Anderson
· Andrei Sakharov

Corresponding to most kinds of particles, there is an associated Antiparticle with the same mass and opposite electric charge. For example, the Antiparticle of the electron is the positively charged antielectron, or positron, which is produced naturally in certain types of radioactive decay.

Feynman diagram

In quantum field theory a Feynman diagram is an intuitive graphical representation of a contribution to the transition amplitude or correlation function of a quantum mechanical or statistical field theory.

Within the canonical formulation of quantum field theory a Feynman diagram represents a term in the Wick's expansion of the perturbative S-matrix. The transition amplitude is the matrix element of the S-matrix between the initial and the final states of the quantum system.

Spin

In quantum mechanics, spin is a fundamental property of atomic nuclei, hadrons, and elementary particles. For particles with non-zero spin, spin direction is an important intrinsic degree of freedom.

As the name indicates, the spin has originally been thought of as a rotation of particles around their own axis.

Stephen William Hawking

Stephen William Hawking CH, CBE, FRS, FRSA is a British theoretical physicist. Hawking is the Lucasian Professor of Mathematics at the University of Cambridge, and a Fellow of Gonville and Caius College, Cambridge. He is known for his contributions to the fields of cosmology and quantum gravity, especially in the context of black holes, and his popular works in which he discusses his own theories and cosmology in general.

Quantization

In physics, Quantization is a procedure for constructing a quantum field theory starting from a classical field theory. This is a generalization of the procedure for building quantum mechanics from classical mechanics. One also speaks of field Quantization, as in the 'Quantization of the electromagnetic field', where one refers to photons as field 'quanta' (for instance as light quanta.)

Angular momentum

Angular momentum is a quantity that is useful in describing the rotational state of a physical system. For a rigid body rotating around an axis of symmetry (e.g. the fins of a ceiling fan), the Angular momentum can be expressed as the product of the body's moment of inertia and its angular velocity ($\mathbf{L} = I\boldsymbol{\omega}$.) In this way, Angular momentum is sometimes described as the rotational analog of linear momentum.

Bosons	In particle physics, Bosons are particles which obey Bose-Einstein statistics; they are named after Satyendra Nath Bose and Albert Einstein. In contrast to fermions, which obey Fermi-Dirac statistics, several Bosons can occupy the same quantum state. Thus, Bosons with the same energy can occupy the same place in space.
Enrico Fermi	Enrico Fermi (29 September 1901 - 28 November 1954) was an Italian physicist most noted for his work on the development of the first nuclear reactor, and for his contributions to the development of quantum theory, nuclear and particle physics, and statistical mechanics. Fermi was awarded the Nobel Prize in Physics in 1938 for his work on induced radioactivity and is today regarded as one of the top scientists of the 20th century. He is acknowledged as a unique physicist who was highly accomplished in both theory and experiment.
Fermion	In particle physics, Fermion s are particles which obey Fermi-Dirac statistics; they are named after Enrico Fermi. In contrast to bosons, which have Bose-Einstein statistics, only one Fermion can occupy a quantum state at a given time; this is the Pauli Exclusion Principle. Thus, if more than one Fermion occupies the same place in space, the properties of each Fermion (e.g. its spin) must be different from the rest.
Wolfgang Ernst Pauli	*Wolfgang Ernst Pauli* (April 25, 1900 - December 15, 1958) was an Austrian theoretical physicist noted for his work on spin theory, and for the discovery of the exclusion principle underpinning the structure of matter and the whole of chemistry.
Symmetry	Symmetry generally conveys two primary meanings. The first is an imprecise sense of harmonious or aesthetically-pleasing proportionality and balance; such that it reflects beauty or perfection. The second meaning is a precise and well-defined concept of balance or 'patterned self-similarity' that can be demonstrated or proved according to the rules of a formal system: by geometry, through physics or otherwise.
Principle	A principle is one of several things: (a) a descriptive comprehensive and fundamental law, doctrine, or assumption; (b) a normative rule or code of conduct, and (c) a law or fact of nature underlying the working of an artificial device. The principle of any effect is the cause that produces it. Depending on the way the cause is understood the basic law governing that cause may acquire some distinction in its expression.
Leptons	Leptons are a family of elementary particles, alongside quarks and gauge bosons (also known as force carriers.) Like quarks, Leptons are fermions (spin-$\frac{1}{2}$ particles) and are subject to the electromagnetic force, the gravitational force, and weak interaction. But unlike quarks, Leptons do not participate in the strong interaction.
Photon	In physics, the photon is the elementary particle responsible for electromagnetic phenomena. It is the carrier of electromagnetic radiation of all wavelengths, including in decreasing order of energy, gamma rays, X-rays, ultraviolet light, visible light, infrared light, microwaves, and radio waves. The photon differs from many other elementary particles, such as the electron and the quark, in that it has zero rest mass; therefore, it travels at the speed of light, c.
Linear	The word Linear comes from the Latin word Linear is, which means created by lines. In mathematics, a Linear map or function f(x) is a function which satisfies the following two properties...

· Additivity (also called the superposition property): f(x + y) = f(x) + f(y.)

Quantum numbers

Quantum numbers describe values of conserved numbers in the dynamics of the quantum system. They often describe specifically the energies of electrons in atoms, but other possibilities include angular momentum, spin etc. Since any quantum system can have one or more quantum numbers, it is a futile job to list all possible quantum numbers.

Electron

An Electron is a subatomic particle that carries a negative electric charge. It has no known substructure and is believed to be a point particle. An Electron has a mass that is approximately 1836 times less than that of the proton.

Neutrinos

Neutrinos are elementary particles that often travel close to the speed of light, lack an electric charge, are able to pass through ordinary matter almost undisturbed and are thus extremely difficult to detect. Neutrinos have a minuscule, but nonzero mass. They are usually denoted by the Greek letter v ' href='/wiki/Nu_(letter)'>nu.)

Radioactivity

Radioactivity can be used in life sciences as a radiolabel to easily visualise components or target molecules in a biological system. Radionuclei are synthesised in particle accelerators and have short half-lives, giving them high maximum theoretical specific activities. This lowers the detection time compared to radionuclei with longer half-lives, such as carbon-14.

Internal energy

In thermodynamics, the Internal energy of a thermodynamic system denoted by U is the total of the kinetic energy due to the motion of molecules (translational, rotational, vibrational) and the potential energy associated with the vibrational and electric energy of atoms within molecules or crystals. It includes the energy in all of the chemical bonds, and the energy of the free, conduction electrons in metals.

One can also calculate the Internal energy of electromagnetic or blackbody radiation.

Potential energy

Potential energy can be thought of as energy stored within a physical system. It is called potential energy because it has the potential to be converted into other forms of energy, such as kinetic energy, and to do work in the process. The standard unit of measure for potential energy is the joule, the same as for work, or energy in general.

Conservation of mass

The law of Conservation of mass matter regardless of the processes acting inside the system. A similar statement is that mass cannot be created/destroyed, although it may be rearranged in space, and changed into different types of particles. This implies that for any chemical process in a closed system, the mass of the reactants must equal the mass of the products.

Physicist

A Physicist is a scientist who studies or practices physics. Physicist s study a wide range of physical phenomena in many branches of physics spanning all length scales: from sub-atomic particles of which all ordinary matter is made (particle physics) to the behavior of the material Universe as a whole (cosmology.)

Most material a student encounters in the undergraduate physics curriculum is based on discoveries and insights of a century or more in the past.

Isolated System

In the natural sciences an Isolated system as contrasted with a open system, is a physical system that does not interact with its surroundings. It obeys a number of conservation laws: its total energy and mass stay constant. They cannot enter or exit, but can only move around inside.

Strangeness

In particle physics, Strangeness S is a property of particles, expressed as a quantum number for describing decay of particles in strong and electro-magnetic reactions, which occur in a short period of time. The Strangeness of a particle is defined as:

$$S = -(n_s - n_{\bar{s}})$$

where $n_{\bar{s}}$ represents the number of strange antiquarks ($\bar{s}$) and n_s represents the number of strange quarks (s.)

The derivation of the phrase 'strange' or 'Strangeness' precedes the discovery of the quark, and was adopted after its discovery in order to preserve the continuity of the phrase; Strangeness of anti-particles being referred to as +1, and particles as −1 as per the original definition.

Parity

In physics, a Parity transformation is the flip in the sign of one spatial coordinate. In three dimensions, it is also commonly described by the simultaneous flip in the sign of all spatial coordinates:

$$P: \begin{pmatrix} x \\ y \\ z \end{pmatrix} \mapsto \begin{pmatrix} -x \\ -y \\ -z \end{pmatrix}$$

A 3×3 matrix representation of P would have determinant equal to -1, and hence cannot reduce to a rotation. The corresponding mathematical notion is that of a point reflection.

Parity transformation

In physics, a parity transformation is the flip in the sign of one spatial coordinate. In three dimensions, it is also commonly described by the simultaneous flip in the sign of all spatial coordinates:

$$P: \begin{pmatrix} x \\ y \\ z \end{pmatrix} \mapsto \begin{pmatrix} -x \\ -y \\ -z \end{pmatrix}$$

A 3×3 matrix representation of P would have determinant equal to -1, and hence cannot reduce to a rotation.

In a two-dimensional plane, parity is not a simultaneous flip of all coordinates, which would be the same as a rotation by 180 degrees.

Chaos

Chaos typically refers to a state lacking order or predictability. In ancient Greece, it referred to the initial state of the universe, and, by extension, space, darkness, or an abyss (the antithetical concept was cosmos.) In modern English, it is used in classical studies with this original meaning; in mathematics and science to refer to a very specific kind of unpredictability; and informally to mean a state of confusion.

Quark	In physics, a quark is a type of subatomic particle. Quarks are elementary fermionic particles which strongly interact due to their color charge. Due to the phenomenon of color confinement, quarks are never found on their own: they are always bound together in composite particles named hadrons.
Thermal neutron	A thermal neutron is a free neutron that is Boltzmann distributed with kT = 0.024 eV (4.0×10^{-21} J) at room temperature. This gives characteristic (not average, or median) speed of 2.2 km/s. The name 'thermal' comes from their energy being that of the room temperature gas or material they are permeating.
Proton	The Proton is a subatomic particle with an electric charge of +1 elementary charge. It is found in the nucleus of each atom but is also stable by itself and has a second identity as the hydrogen ion, $^1H^+$. It is composed of three even more fundamental particles comprising two up quarks and one down quark.
Rutherford	The Rutherford is an obsolete unit of radioactivity, defined as the activity of a quantity of radioactive material in which one million nuclei decay per second. It is therefore equivalent to one megabecquerel. It was named after Ernest Rutherford It is not an SI unit.
Glueball	In particle physics, a Glueball is a hypothetical composite particle. It solely consists of gluon particles, without valence quarks. Such a state is possible because gluons carry color charge and experience the strong interaction.
Quantum chromodynamics	Quantum chromodynamics is a theory of the strong interaction, a fundamental force describing the interactions of the quarks and gluons found in hadrons. It is the study of the S Yang-Mills theory of color-charged fermions. QCD is a quantum field theory of a special kind called a non-abelian gauge theory.
Electric charge	Electric charge is a fundamental conserved property of some subatomic particles, which determines their electromagnetic interaction. Electrically charged matter is influenced by, and produces, electromagnetic fields. The interaction between a moving charge and an electromagnetic field is the source of the electromagnetic force, which is one of the four fundamental forces.
Top quark	The top quark is the third-generation up-type quark with a charge of e. It was discovered in 1995 by the CDF and D0 experiments at Fermilab, and is by far the most massive of known elementary particles. Its mass is measured at 172.6 ± 1.4 GeV/c^2, about the same weight as the nuclei of tantalum or tungsten atoms.
Upsilon	Upsilon is the 20th letter of the Greek alphabet. In the system of Greek numerals it has a value of 400. It is derived from the Phoenician waw.
Higgs mechanism	In the standard model of particle physics, the Higgs mechanism is a theoretical framework which explains how the masses of the W and Z bosons arise as a result of electroweak symmetry breaking. More generally, the Higgs mechanism is the way that the gauge bosons in any gauge theory, like the standard model, get a nonzero mass. It requires an extra field, a Higgs field, which interacts with the gauge fields, and which has a nonzero value in its lowest energy state, a vacuum expectation value.

Particle accelerator	A particle accelerator is a device that uses electric fields to propel electrically-charged particles to high speeds and to contain them. An ordinary CRT television set is a simple form of accelerator. There are two basic types: linear accelerators and circular accelerators.
Universe	The Universe is defined as everything that physically exists: the entirety of space and time, all forms of matter, energy and momentum, and the physical laws and constants that govern them. However, the term 'universe' may be used in slightly different contextual senses, denoting such concepts as the cosmos, the world or Nature. Astronomical observations indicate that the universe is 13.73 ± 0.12 billion years old and at least 93 billion light years across.
Cosmological constant	In physical cosmology, the Cosmological constant was proposed by Albert Einstein as a modification of his original theory of general relativity to achieve a stationary universe. Einstein abandoned the concept after the observation of the Hubble redshift indicated that the universe might not be stationary, as he had based his theory on the idea that the universe is unchanging. However, the discovery of cosmic acceleration in the 1990s has renewed interest in a Cosmological constant
General relativity	General relativity or the general theory of relativity is the geometric theory of gravitation published by Albert Einstein in 1916. It is the current description of gravitation in modern physics. It unifies special relativity and Newton's law of universal gravitation, and describes gravity as a geometric property of space and time, or spacetime.
Supernova	A supernova is a stellar explosion. They are extremely luminous and cause a burst of radiation that often briefly outshines an entire galaxy before fading from view over several weeks or months. During this short interval, a supernova can radiate as much energy as the Sun could emit over its life span.
Telescope	A telescope is an instrument designed for the observation of remote objects by the collection of electromagnetic radiation. The first known practically functioning telescope s were invented in the Netherlands at the beginning of the 17th century. ' telescope s' can refer to a whole range of instruments operating in most regions of the electromagnetic spectrum.
Optics	Optics is the science that describes the behavior and properties of light and the interaction of light with matter. Optics explains optical phenomena. The word optics comes from á½€πτικÎ®, meaning appearance or look in Ancient Greek.)
Radiation	Radiation, as in physics, is energy in the form of waves or moving subatomic particles emitted by an atom or other body as it changes from a higher energy state to a lower energy state. Radiation can be classified as ionizing or non-ionizing radiation, depending on its effect on atomic matter. The most common use of the word 'radiation' refers to ionizing radiation.
Acceleration	In physics, and more specifically kinematics, Acceleration is the change in velocity over time. Because velocity is a vector, it can change in two ways: a change in magnitude and/or a change in direction. In one dimension, Acceleration is the rate at which something speeds up or slows down.

Spacetime

In physics, spacetime is any mathematical model that combines space and time into a single construct called the spacetime continuum. Spacetime is usually interpreted with space being three-dimensional and time playing the role of the fourth dimension. According to Euclidean space perception, the universe has three dimensions of space and one dimension of time.

Space

Space is the boundless, three-dimensional extent in which objects and events occur and have relative position and direction. Physical Space is often conceived in three linear dimensions, although modern physicists usually consider it, with time, to be part of the boundless four-dimensional continuum known as Space time. In mathematics Space s with different numbers of dimensions and with different underlying structures can be examined.

Gravitational lens

A Gravitational lens is formed when the light from a very distant, bright source (such as a quasar) is 'bent' around a massive object (such as a cluster of galaxies) between the source object and the observer. The process is known as Gravitational lens ing, and is one of the predictions of Albert Einstein's general theory of relativity

Orbit

In physics, an orbit is the gravitationally curved path of one object around a point or another body, for example the gravitational orbit of a planet around a star.

Historically, the apparent motion of the planets were first understood in terms of epicycles, which are the sums of numerous circular motions. This predicted the path of the planets quite well, until Johannes Kepler was able to show that the motion of the planets were in fact elliptical motions.

Orbits

The velocity relationship of two objects with mass can thus be considered in four practical classes, with subtypes:

- No orbit
- Interrupted orbits

- Range of interrupted elliptical paths
- Circumnavigating orbits

- Range of elliptical paths with closest point opposite firing point
- Circular path
- Range of elliptical paths with closest point at firing point
- Infinite orbits

- Parabolic paths
- Hyperbolic paths

In many situations relativistic effects can be neglected, and Newton's laws give a highly accurate description of the motion. Then the acceleration of each body is equal to the sum of the gravitational forces on it, divided by its mass, and the gravitational force between each pair of bodies is proportional to the product of their masses and decreases inversely with the square of the distance between them. To this Newtonian approximation, for a system of two point masses or spherical bodies, only influenced by their mutual gravitation, the orbits can be exactly calculated. If the heavier body is much more massive than the smaller, as for a satellite or small moon orbiting a planet or for the Earth orbiting the Sun, it is accurate and convenient to describe the motion in a coordinate system that is centered on the heavier body, and we can say that the lighter body is in orbit around the heavier.

Russell Alan Hulse

Russell Alan Hulse is an American physicist and winner of the Nobel Prize in Physics, shared with his thesis advisor Joseph Hooton Taylor Jr., 'for the discovery of a new type of pulsar, a discovery that has opened up new possibilities for the study of gravitation'. He was a specialist in the pulsar studies and gravitational waves.

Hulse was born in New York City and attended Bronx High School of Science and the Cooper Union before moving to University of Massachusetts Amherst (Ph.D. Physics 1975.)

Precession

Precession refers to a change in the direction of the axis of a rotating object. In physics, there are two types of precession, torque-free and torque-induced, the latter being discussed here in more detail. In certain contexts, 'precession' may refer to the precession that the Earth experiences, the effects of this type of precession on astronomical observation, or to the precession of orbital objects..

Pulsars

Pulsars are highly magnetized rotating neutron stars that emit a beam of electromagnetic radiation in the form of radio waves. Their observed periods range from 1.4 ms to 8.5 s. The radiation can only be observed when the beam of emission is pointing towards the Earth.

Euclid

Euclid , fl. 300 BC, also known as Euclid of Alexandria, was a Greek mathematician and is often referred to as the 'Father of Geometry.' He was active in Hellenistic Alexandria during the reign of Ptolemy I . His Elements is the most successful textbook and one of the most influential works in the history of mathematics, serving as the main textbook for teaching mathematics (especially geometry) from the time of its publication until the late 19th or early 20th century.

Flatness problem

The Flatness problem is a cosmological fine-tuning problem within the Big Bang model of the universe. Such problems arise from observation that some of the initial conditions of the universe appear to fine-tuned to very 'special' values, and that a small deviation from these values would have had massive effects on the nature of the universe at the current time.

In the case of the Flatness problem the parameter which appears fine-tuned is the density of matter and energy in the universe.

Axion

The Axion is a hypothetical elementary particle postulated by the Peccei-Quinn theory in 1977 to resolve the strong-CP problem in quantum chromodynamics (QCD.) In 2005, an experimental search by the PVLAS collaboration reported results suggesting Axion detection, but new experiments performed by the PVLAS team exclude this result. The 2005 PVLAS results were problematic because compatibility with the negative results of other searches, such as CAST, as well as astrophysical limits, ruled out standard Axion scenarios, while alternative hypotheses have been postulated by other researchers.

Dark energy

In physical cosmology, astronomy and celestial mechanics, Dark energy is a hypothetical form of energy that permeates all of space and tends to increase the rate of expansion of the universe. Dark energy is the most popular way to explain recent observations that the universe appears to be expanding at an accelerating rate. In the standard model of cosmology, Dark energy currently accounts for 74% of the total mass-energy of the universe.

Neutralino

In particle physics, the Neutralino is a hypothetical particle, part of the doubling of the menagerie of particles predicted by supersymmetric theories. The standard symbol for Neutralino s is $\tilde{\chi}_i^0$ (chi), where i runs from 1 to 4.

In supersymmetry models, all Standard Model particles have partner particles with the same quantum numbers except for the quantum number spin, which differs by 1/2 from its partner particle.

Riemannian geometry

Riemannian geometry is the branch of differential geometry that studies Riemannian manifolds, smooth manifolds with a Riemannian metric. This gives in particular local notions of angle, length of curves, surface area, and volume. From those some other global quantities can be derived by integrating local contributions.

Adaptive optics

Adaptive optics is a technology used to improve the performance of optical systems by reducing the effects of rapidly changing optical distortion. It is used in astronomical telescopes and laser communication systems to remove the effects of atmospheric distortion, and in retinal imaging systems to reduce the impact of ocular aberrations. Adaptive optics works by measuring the distortions in a wavefront and compensating for them with a spatial phase modulator such as a deformable mirror or liquid crystal array.

Dark matter

In astronomy and cosmology, Dark matter is hypothetical matter that is undetectable by its emitted radiation, but whose presence can be inferred from gravitational effects on visible matter. According to present observations of structures larger than galaxies, as well as Big Bang cosmology, Dark matter and dark energy could account for the vast majority of the mass in the observable universe. Dark matter is postulated to partially account for evidence of 'missing mass' in the universe, including the rotational speeds of galaxies, orbital velocities of galaxies in clusters, gravitational lensing of background objects by galaxy clusters such as the Bullet Cluster, and the temperature distribution of hot gas in galaxies and clusters of galaxies.

Density

The Density of a material is defined as its mass per unit volume. The symbol of Density is ρ '>rho.)

Mathematically:

$$\rho = \frac{m}{V}$$

where:

ρ (rho) is the Density
m is the mass,
V is the volume.

M-Theory

In theoretical physics, M-theory is an extension of string theory in which 11 dimensions are identified. Because the dimensionality exceeds the dimensionality of five superstring theories in 10 dimensions, it is believed that the 11-dimensional theory unifies all string theories (and supersedes them.) Though a full description of the theory is not yet known, the low-entropy dynamics are known to be supergravity interacting with 2- and 5-dimensional membranes.

Matter

The term Matter traditionally refers to the substance that objects are made of. One common way to identify this 'substance' is through its physical properties; a common definition of Matter is anything that has mass and occupies a volume. However, this definition has to be revised in light of quantum mechanics, where the concept of 'having mass', and 'occupying space' are not as well-defined as in everyday life.

Superstring theory

Superstring theory is an attempt to explain all of the particles and fundamental forces of nature in one theory by modelling them as vibrations of tiny supersymmetric strings. It is considered one of the most promising candidate theories of quantum gravity. Superstring theory is a shorthand for supersymmetric string theory because unlike bosonic string theory, it is the version of string theory that incorporates fermions and supersymmetry.

False vacuum

In quantum field theory, a False vacuum is a metastable sector of space which appears to be a perturbative vacuum but is unstable to instanton effects which may tunnel to a lower energy state. This tunneling can be caused by quantum fluctuations or the creation of high energy particles. Simply put, the False vacuum is a local minimum, but not the lowest energy state, even though it may remain stable for some time.

Phase

A phase is one part or portion in recurring or serial activities or occurrences logically connected within a greater process, often resulting in an output or a change. Phase or phases may also refer to:

· Phase, a physically distinctive form of a substance, such as the solid, liquid, and gaseous states of ordinary matter
· Phase transition is the transformation of a thermodynamic system from one phase to another
· The initial condition of a cyclic phenomenon

· Phase, initial angle of a sinusoid function at its origin
· Continuous Fourier transform, angle of a complex coefficient representing the phase of one sinusoidal component
· The current state of a cyclic phenomenon

· Lunar phase, the appearance of the Moon as viewed from the Earth
· Planetary phase, the appearance of the illuminated section of a planet
· Instantaneous phase, generalization for both cyclic and non-cyclic phenomena
· Phase factor, a complex scalar in the context of quantum mechanics
· Polyphase system, a means of distributing alternating current electric power in multiple conducting wires with definite phase offsets

· Single-phase electric power
· Three-phase electric power Basics of three-phase electric power
· Three-phase Mathematics of three-phase electric power
· In biology, a part of the cell cycle in which cells divide and reproduce
· Phaser, an audio effect.
· Archaeological phase, a discrete period of occupation at an archaeological site.

- 'Phases', an episode of the TV series Buffy the Vampire Slayer
- Phases, fictional boss monsters from the .hack franchise
- Phase, an incarnation of the DC Comics character usually known as Phantom Girl
- Phase IV, a 1974 science fiction movie directed by Saul Bass

- A phase is a musical composition using Steve Reich's phasing technique.
- Phase, a syntactic domain hypothesized by Noam Chomsky
- Phase 10, a card game created by Fundex Games
- Phase, a music game for the iPod developed by Harmonix Music Systems
- Phase usually a period of combat within a larger military operation .

Phase transition

In thermodynamics, phase transition or phase change is the transformation of a thermodynamic system from one phase to another. The distinguishing characteristic of a phase transition is an abrupt change in one or more physical properties, in particular the heat capacity, with a small change in a thermodynamic variable such as the temperature.

In the English vernacular, the term is most commonly used to describe transitions between solid, liquid and gaseous states of matter, in rare cases including plasma.

Vacuum

A vacuum is a volume of space that is essentially empty of matter, such that its gaseous pressure is much less than atmospheric pressure. The word comes from the Latin term for 'empty,' but in reality, no volume of space can ever be perfectly empty. A perfect vacuum with a gaseous pressure of absolute zero is a philosophical concept that is never observed in practice.

Temperature

Temperature is a physical property of a system that underlies the common notions of hot and cold; something that is hotter generally has the greater temperature. Specifically, temperature is a property of matter. Temperature is one of the principal parameters of thermodynamics.

Transition temperature

Transition temperature is the temperature at which material changes from one crystal to another. There are total seven crystal systems and every material is known to exist in one of them. At any critical value of temperature it can shift to any crystal system from the previously existing one.

CPSIA information can be obtained at www.ICGtesting.com
Printed in the USA
BVOW080938090212

282454BV00001B/245/P

9 781428 897823